Lecture Notes in Mathematics 1735

Springer
Berlin
Heidelberg
New York
Barcelona
Hong Kong
London
Milan
Paris
Singapore
Tokyo

Dmitri Yafaev

Scattering Theory: Some Old and New Problems

Springer

Author

Dmitri R. Yafaev
Department of Mathematics
University of Rennes-1
Campus de Beaulieu
35042 Rennes, France

E-mail: yafaev@univ-rennes1.fr

Cataloging-in-Publication Data applied for

Die Deutsche Bibliothek - CIP-Einheitsaufnahme

Jafaev, Dmitrij R.:
Scattering theory : some old and new problems / Dimitri Yafaev. -
Berlin ; Heidelberg ; New York ; Barcelona ; Hong Kong ; London ;
Milan ; Paris ; Singapore ; Tokyo : Springer, 2000
(Lecture notes in mathematics ; 1735)
ISBN 3-540-67587-6

Mathematics Subject Classification (2000): 35P25, 47A40, 81U05, 81U10

ISSN 0075-8434
ISBN 3-540-67587-6 Springer-Verlag Berlin Heidelberg New York

Springer-Verlag is a company in the BertelsmannSpringer publishing group.

Printed in Germany

Typesetting: Camera-ready TEX output by the author
Printed on acid-free paper SPIN: 10725026 41/3143/du 543210

Introduction

1. Scattering theory has its origin in quantum mechanics and is intimately related to the theory of partial differential equations. From mathematical point of view it can be considered as perturbation theory of self-adjoint operators on the (absolutely) continuous spectrum. In general, perturbation theory draws conclusions about a self-adjoint operator H (acting in a Hilbert space $\mathcal{H}$) given an information regarding a simpler operator H_0. Thereby it is required that the operators H_0 and H be close in a sense depending on a particular problem. In physical terms the operator H_0 (the free Hamiltonian) describes a system of non-interacting particles (or clusters of particles), while the "full" Hamiltonian H describes the real system including interactions.

Scattering theory is concerned with a study of the behaviour for large times of solutions of the time-dependent equation $i\partial u/\partial t = Hu$ in terms of the free equation $i\partial u_0/\partial t = H_0 u_0$. It turns out that under appropriate assumptions on the perturbation $V = H - H_0$, for every f orthogonal to eigenvectors of H, there exists $f_0^\pm$ such that

$$\lim_{t\to\pm\infty} ||u(t) - u_0(t)|| = 0,$$

if $u(0) = f$ and $u_0(0) = f_0^\pm$. Since $u(t) = \exp(-iHt)f$ and $u_0(t) = \exp(-iHt)f_0^\pm$, the initial data f and $f_0^\pm$ are related by the equality

$$f = \lim_{t\to\pm\infty} \exp(iHt)\exp(-iH_0t)f_0^\pm.$$

The mapping $W^\pm = W^\pm(H, H_0) : f_0^\pm \mapsto f$ is known as the wave operator. By definition, $W^\pm$ exists if the limit relating $f_0^\pm$ and f exists for any $f_0^\pm$ (provided H_0 is absolutely continuous). Wave operators are automatically isometric. Moreover, under the assumption of their existence, the wave operators enjoy the intertwining property $HW^\pm = W^\pm H_0$. Therefore the range of $W^\pm$ belongs to the absolutely continuous subspace of H. The operator $W^\pm$ is called complete if its range coincides with this subspace. Then the free Hamiltonian H_0 and the absolutely continuous part of H are unitarily equivalent. In physical applications Hamiltonians do not usually have the singular continuous spectrum. If H_0 is not absolutely continuous, then the above definition of $W^\pm$ works for $f_0^\pm$ from the absolutely continuous subspace $\mathcal{H}_0^{(ac)}$ of H_0; then $W^\pm$ is extended by zero to $\mathcal{H} \ominus \mathcal{H}_0^{(ac)}$.

If H_0 and H are not close enough, then it is sometimes still possible to prove the existence of more general wave operators

$$W^{\pm} = W^{\pm}(H, H_0; J) = s - \lim_{t\to\pm\infty} \exp(iHt) J \exp(-iH_0 t) P_0,$$

where the "identification" J is a bounded operator, P_0 is the orthogonal projection on $\mathcal{H}_0^{(ac)}$ and the limit is strong. These limits exist if the "effective perturbation" $HJ - JH_0$ is in some sense small. The intertwining property remains true for wave operators $W^{\pm}(H, H_0; J)$, but their isometricity may be lost.

Another important object is the scattering operator $\mathbf{S} = (W^+)^* W^-$ which connects directly the asymptotic behaviour of a quantum system as $t \to -\infty$ and $t \to \infty$. Its consideration allows one to avoid a study of the evolution of a system for finite times. Since the operator $\mathbf{S}$ commutes with H_0, it reduces to multiplication by an operator-valued function $S(\lambda)$, known as the scattering matrix, in a diagonal representation of the free operator H_0. The operators $S(\lambda)$ are unitary for (almost) all λ if both wave operators $W^{\pm}$ exist, are isometric and complete.

We use consistently the smooth method (see, e.g., [134], v.3, or [165], for a detailed presentation) for the proof of the existence and completeness of wave operators. Recall that an H-bounded operator K is called H-smooth (in the sense of T. Kato) if

$$\int_{-\infty}^{\infty} ||K \exp(-iHt) f||^2 dt \le C||f||^2.$$

This notion can be reformulated in an equivalent way in terms of the resolvent $R(z) = (H - z)^{-1}$, $\operatorname{Im} z \ne 0$, as uniform boundedness of the operator-function $K(R(z) - R(\bar{z})K^*$. The notion of H-smoothness can be localized on any spectral interval Λ of H which permits to avoid eigenvalues and other exceptional points of H. Namely, K is H-smooth on Λ if the operator $KE(\Lambda)$ is H-smooth (E and E_0 are the spectral families of the operators H and H_0).

Suppose now that $HJ - JH_0 = K^* K_0$ where K_0 and K are, respectively, H_0 - and H-smooth on a set of intervals Λ exhausting $\mathbb{R}$ up to a set of Lebesgue measure zero. Then

$$(Je^{-iH_0 t} f_0, e^{-iHt} f) = (Jf_0, f) + i \int_0^t (K_0 e^{-iH_0 s} f_0, K e^{-iHs} f) ds,$$

where the integral is convergent for $f_0 = E_0(\Lambda) f_0$ and $f = E(\Lambda) f$ uniformly either in f_0 for $||f_0|| \le 1$ or in f for $||f|| \le 1$. This implies the existence of the wave operators $W^{\pm}(H, H_0; J)$ and $W^{\pm}(H_0, H; J^*)$. In particular, in the case $J = I$ we have that $W^{\pm}(H, H_0)$ are complete.

2. The Schrödinger operator $H = -\Delta + V(x)$ in the space $\mathcal{H} = L_2(\mathbb{R}^d)$ with a potential V decaying at infinity is a typical object of scattering theory. More general differential operators whose coefficients have limits as $|x| \to \infty$ can be treated essentially similarly. The operator H describes two interacting particles which may be either in a bound state or asymptotically (as the time $t \to \infty$ or $t \to -\infty$) free. This statement is called asymptotic completeness. If $|V(x)| \le C(1 + |x|)^{-\rho}$,

$\rho > 1$ (the short-range case), then the kinetic energy operator $H_0 = -\Delta$ plays the role of the unperturbed operator. In this case the wave operators $W^{\pm}$ exist and are complete. For the Schrödinger operator, as well as for more general differential operators, only the completeness of wave operators is a substantial mathematical problem.

A relatively simple proof of this result can be obtained within the framework of the smooth method. Let $\langle x \rangle$ be multiplication by $\langle x \rangle = (1 + |x|^2)^{1/2}$. One of the fundamental analytical results of scattering theory is the H-smoothness of the operator $\langle x \rangle^{-l}$ for any $l > 1/2$ on any compact interval $\Lambda \subset (0, \infty)$. This result is known as the limiting absorption principle. For the free operator $H_0 = -\Delta$ it follows directly from the Sobolev trace theorem $\mathsf{H}^l(\mathbf{R}^d) \subsetneq L_2(\mathbb{S}^{d-1})$ where $l > 1/2$. Then the limiting absorption principle is deduced for the Schrödinger operator H with the help of the resolvent identity

$$R(z) = R_0(z) - R_0(z)VR(z), \quad R_0(z) = (H_0 - z)^{-1}.$$

As explained in the previous subsection, this implies the existence of both wave operators $W^{\pm}(H, H_0)$ and $W^{\pm}(H_0, H)$ and hence the completeness of $W^{\pm}(H, H_0)$.

The operator H_0 reduces by the Fourier transform to multiplication by the independent variable λ (which plays the role of the energy) in the space $L_2(\mathbf{R}_+; L_2(\mathbb{S}^{d-1}))$. Therefore the scattering matrix $S(\lambda)$ for the pair H_0, $H = -\Delta + V$ acts in the space $L_2(\mathbb{S}^{d-1})$ and is a unitary operator for all $\lambda > 0$. Moreover, the operator $S(\lambda) - I$ is compact, and hence the spectrum of $S(\lambda)$ consists of eigenvalues of finite multiplicity (except, possibly, the eigenvalue 1) lying on the unit circle and accumulating at the point 1 only.

The scattering matrix can also be defined in terms of solutions of the stationary Schrödinger equation

$$-\Delta\psi + V(x)\psi = \lambda\psi.$$

If $V(x) = O(|x|^{-\rho})$ with $\rho > (d+1)/2$, then, for any $\lambda > 0$ and any unit vector $\omega \in \mathbb{S}^{d-1}$, this equation has a (unique) solution with the asymptotics

$$\exp(i\lambda^{1/2}\langle \omega, x \rangle) + a|x|^{-(d-1)/2} \exp(i\lambda^{1/2}|x|)$$

as $|x| \to \infty$. The coefficient $a = a(\hat{x}, \omega; \lambda)$ depends on the incident direction ω of the incoming plane wave $\exp(i\lambda^{1/2}\langle \omega, x \rangle)$, its energy λ and the direction $\hat{x} = x|x|^{-1}$ of observation of the outgoing spherical wave $|x|^{-(d-1)/2} \exp(i\lambda^{1/2}|x|)$. The function $a(\theta, \omega; \lambda)$ is called the scattering amplitude, and $S(\lambda) - I$ is the integral operator with kernel $ie^{\pi i(d-3)/4}\lambda^{(d-1)/4}(2\pi)^{-(d-1)/2}a(\theta, \omega; \lambda)$.

From the point of view of quantum mechanics the plane wave describes a beam of particles incident on a scattering center, and the outgoing spherical wave corresponds to scattered particles. The ratio of the flux density of the scattered particles to that of the incident beam is $|a(\theta, \omega; \lambda)|^2 d\theta$. This quantity is the main observable in scattering experiments. It is called the scattering cross-section in the solid angle $d\theta$. Further details on the quantum mechanical picture of scattering may be found in the textbooks [112] and [51].

3. Potentials decaying slower than (or as) the Coulomb potential $|x|^{-1}$ at infinity are called long-range. More precisely, it is required that $|\partial^\alpha V(x)| \leq C(1+|x|)^{-\rho-|\alpha|}$, $\rho \in (0,1]$, for all derivatives of V up to some order. In the long-range case the wave operators $W^\pm(H, H_0)$ do not exist, and the asymptotic dynamics should be properly modified. It can be done in a time-dependent way either in the coordinate or momentum representations. For example, in the coordinate representation the free evolution $\exp(-iH_0t)$ should be replaced in the defiition of wave operators by unitary operators $U_0(t)$ defined by

$$(U_0(t)f)(x) = \exp(i\Xi(x,t))(2it)^{-d/2}\hat{f}(x/(2t)),$$

where $\hat{f}$ is the Fourier transform of f. For short-range potentials we can set $\Xi(x,t) = (4t)^{-1}|x|^2$. In the long-range case the phase function $\Xi(x,t)$ should be chosen as a (perhaps, approximate) solution of the eikonal equation

$$\partial\Xi/\partial t + |\nabla\Xi|^2 + V = 0.$$

Thus both in short- and long-range cases solutions of the time-dependent Schrödinger equation "live" in a region of the configuration space where $|x|$ is of order $|t|$. Long-range potentials change only phases of these solutions.

Another possibility is a time-independent modification in the phase space. Actually, we consider wave operators $W^\pm(H, H_0; J)$ where J is a pseudo-differential operator with oscillating symbol $\exp(i\Phi(x,\xi))\zeta(x,\xi)$. Due to the conservation of energy, we may suppose that $\zeta(x,\xi)$ contains a factor $\psi(|\xi|^2)$ with $\psi \in C_0^\infty(0,\infty)$. Set $\varphi(x,\xi) = \langle x, \xi\rangle + \Phi(x,\xi)$. The perturbation $HJ - JH_0$ is also a pseudo-differential operator, and its symbol is short-range (it is $O(|x|^{-1-\varepsilon})$, $\varepsilon > 0$, as $|x| \to \infty$) if $\exp(i\varphi(x,\xi))\zeta(x,\xi)$ is an approximate eigenfunction of the operator H corresponding to the "eigenvalue" $|\xi|^2$. This leads to the eikonal equation $|\nabla_x\varphi(x,\xi)|^2 + V(x) = |\xi|^2$. The notorious difficulty (for $d \geq 2$) of this approach is that the eikonal equation does not have (even approximate) solutions such that $|\nabla_x\Phi(x,\xi)| \to 0$ as $|x| \to \infty$ and the arising error term is short-range. However it is easy to construct functions $\varphi = \varphi^\pm$ (for example,

$$\Phi^\pm(x,\xi) = \pm 2^{-1}\int_0^\infty \Big(V(x \pm \tau\xi) - V(\pm\tau\xi)\Big)d\tau$$

for $\rho > 1/2$) satisfying these conditions if a conical neighbourhood of the direction $\mp\xi$ is removed from $\mathbb{R}^d$. This amounts to the choice of $\zeta(x,\xi) = \zeta^\pm(x,\xi)$ as a cut-off function which is homogeneous function of order zero in the variable x. We emphasize that now we have a couple of different identifications $J = J^\pm$.

The long-range problem is essentially more difficult than the short-range one. The limiting absorption principle remains true in this case, but its proof cannot be performed within perturbation theory. The simplest proof relies on the celebrated Mourre estimate for the commutator $i[H, \mathbf{A}]$ of H with the generator of dilations $\mathbf{A}$. This estimate affirms that the operator $iE(\Lambda_\lambda)[H, \mathbf{A}]E(\Lambda_\lambda)$ is positively definite if Λ_λ is a sufficiently small neighbourhood of a point $\lambda > 0$. The H-smoothness of the operator $\langle x\rangle^{-l}$, $l > 1/2$, is deduced from this fact by some arguments of abstract nature (they do not really use concrete forms of the operators H and $\mathbf{A}$).

However the limiting absorption principle is not sufficient for construction of scattering theory in the long-range case. We argue that, for the proof of completeness of modified wave operators, it should be supplemented by one additional estimate. To formulate it, denote by $(\nabla^{\perp}u)(x)$ the orthonal projection of a vector $(\nabla u)(x)$ on the plane orthogonal to x. Then the operator $G = \langle x\rangle^{-1/2}\nabla^{\perp}$ is H-smooth on any compact $\Lambda \subset (0,\infty)$. This result is formulated as an estimate (either on the resolvent or on the unitary group of H) which we refer to as the radiation estimate. This estimate is not very astonishing from the viewpoint of analogy with the classical mechanics. Indeed, in the case of free motion, the vector $x(t)$ of the position of a particle is directed asymptotically as its momentum ξ. Regarded as a pseudo-differential operator, $\nabla^{\perp}$ has a symbol which equals zero if $x = \gamma\xi$ for some $\gamma \in \mathbb{R}$. Thus $\nabla^{\perp}$ removes the part of the phase space where a classical particle propagates. Our proof of the radiation estimate is based on the inequality

$$G^*G \leq C_0[H,\partial_r] + C_1\langle x\rangle^{-1-\rho}, \quad \partial_r = \partial/\partial|x|,$$

which can be obtained by a direct calculation. Since the integral

$$i\int_0^t([H,\partial_r]e^{-iHs}f, e^{-iHs})f)ds = (\partial_r e^{-iHt}f, e^{-iHt}f) - (\partial_r f, f)$$

is bounded by $C(\Lambda)||f||^2$ for $f \in E(\Lambda)f$ and the operator $\langle x\rangle^{-(1+\rho)/2}$ is H-smooth on Λ, this implies H-smoothness of the operator $GE(\Lambda)$.

Calculating the perturbation $HJ^{\pm} - J^{\pm}H_0$ we see that it is a sum of two pseudo-differential operators. The first of them is short-range and thus can be taken into account by the limiting absorption principle. The symbol of the second one contains the first derivatives (in the variable x) of the cut-off function $\zeta^{\pm}(x,\xi)$ and hence decreases at infinity as $|x|^{-1}$ only. This operator factorizes into a product of H_0- and H-smooth operators according to the radiation estimate. Thus all wave operators $W^{\pm}(H,H_0;J^{\pm})$ and $W^{\pm}(H_0,H;(J^{\pm})^*)$ exist. These operators are isometric since the operators $J^{\pm}$ are in some sense close to unitary operators. The isometricity of $W^{\pm}(H_0,H;(J^{\pm})^*)$ is equivalent to the completeness of $W^{\pm}(H,H_0;J^{\pm})$.

Although the modified wave operators enjoy basically the same properties as in the short-range case, properties of the scattering matrices in the short- and long-range cases are drastically different. Here we note only that for long-range potentials, due to a wild diagonal singularity of kernel of the scattering matrix, its spectrum covers the whole unit circle.

4. Another important problem of scattering theory concerns the Schrödinger operator of N interacting particles. Its qualitative difference compared to the two-particle case is that the potential energy of pair interactions $V^{(ij)}(x_i - x_j)$, $i,j = 1,\cdots,N$, $i \neq j$, (x_i is a position of the particle labelled by i) between particles does not tend to zero at infinity in the configuration space of a system, even if the center of mass motion is removed. Therefore the formulation of the multiparticle scattering problem takes into account all possible break-ups $a = \{C_1, \dots, C_p\}$ of a system of particles into clusters $C_1, \dots, C_p$, $2 \le p \le N$. For large times, particles of the same cluster C_l, $l = 1, \dots, p$, form a bound state, and different clusters do not interact with each other if all pair potentials $V^{(ij)}$ are short-range. The asymptotic evolution determined by clusters $C_1, \dots, C_p$ and bound states of all these clusters is called scattering channel. Physically it is natural to expect that the list of all such channels is exhaustive, that is no other scattering process is possible. This statement is called asymptotic completeness. The kinetic energy operator $H_0 = -\Delta$ describes only one channel of scattering where all N particles are asymptotically free. Other channels of scattering are described by the Hamiltonians

$$H_a = H_0 + V^a, \quad V^a = \sum_{l=1}^{p} \sum_{i,j \in C_l, i<j} V^{(ij)},$$

which contain only interactions of particles from the same cluster.

This intuitive picture of scattering is essentially preserved for long-range pair potentials $V^{(ij)}$ decaying faster than $|x_i - x_j|^{-\rho}$ as $|x_i - x_j| \to \infty$ for some $\rho > \sqrt{3} - 1$. On the contrary, if $\rho < 1/2$ new channels of scattering might appear in a system of N, $N \ge 3$, particles.

The limiting absorption principle remains true for the N-particle Schrödinger operator H (even with long-range pair potentials), that is the operator $\langle x \rangle^{-l}$ is H-smooth for any $l > 1/2$ if some closed countable set Υ (of eigenvalues and thresholds of H) is removed from the real axis. Its only available proof relies again on the Mourre estimate. If the limiting absorption principle were true for the critical value $l = 1/2$, then it would have implied asymptotic completeness for an arbitrary N. Unfortunately the operator $\langle x \rangle^{-1/2}$ is definitely not smooth even with respect to the free operator H_0.

As in the two-particle long-range problem, the proof of asymptotic completeness for multiparticle systems requires additionally a radiation estimate. However, in the N-particle case this estimate looks different in different regions of the configuration space. It is the same as for the two-particle case in the "free" region where all particles are far from each other. In the general case one needs to introduce coordinates x^a, x_a depending on the break-up a of N particles into clusters $C_1, \dots, C_p$. Namely, x^a is the set of coordinates of particles with respect to the centers of mass of the corresponding clusters, and x_a describes relative positions of centers of masses of all clusters. Let $\chi_a(x)$ be the characteristic function of the region where particles from each cluster C_l, $l = 1, \dots, p$, are close to each other compared to distances between different clusters. Then the operator $G_a = \chi_a \langle x \rangle^{-1/2} \nabla_{x_a}^{\perp}$ is H-smooth on any compact interval Λ such that $\Lambda \cap \Upsilon = \emptyset$. This is of course natural since in this region the motion of a system is expected to be asymptotically free in the variable x_a. On the contrary, this motion is very complicated in the variable x^a pertaining to bound states of different clusters.

As in the two-particle case, our proof of the radiation estimates is based on consideration of the commutator of H with some differential operator $M = \sum (m^{(j)} D_j +$

$D_j m^{(j)})$ where $m^{(j)} = \partial m/\partial x_j$. Here m is a specially constructed function satisfying the following properties: 1^0 $m(x)$ is homogeneous of order 1; 2^0 for any cluster decomposition b it does not depend on x^b in some conical neighbourhood of the subspace $x^b = 0$; 3^0 $m(x)$ is convex; 4^0 $m(x) = \mu_a |x_a|$, $\mu_a \geq 1$, on support of the function χ_a. Due to the properties 1^0 and 2^0 the commutator $[V, M]$ is a short-range function (estimated by $\langle x \rangle^{-1-\varepsilon}$ for $\varepsilon > 0$). Due to the properties 3^0 and 4^0 the commutator $[H_0, M] \geq c G_a^* G_a$, $c > 0$, up to short-range terms. Similarly to the two-particle case, this implies that the operators G_a are H-smooth on Λ.

Given the limiting absorption principle and the radiation estimates, we first check the existence of auxiliary wave operators

$$W^{\pm}(H, H_a; M_a E_a(\Lambda)) \quad \text{and} \quad W^{\pm}(H_a, H; M_a E(\Lambda))$$

where $M_a = \sum(m_a^{(j)} D_j + D_j m_a^{(j)})$ are again differential operators with coefficients $m_a^{(j)} = \partial m_a/\partial x_j$. The functions m_a satisfy the properties 1^0, 2^0 formulated above and $m^a(x) = 0$ in some conical neighbourhoods of the subspaces $x_i = x_j$ for all i, j belonging to different clusters $C_1, \ldots, C_p$. Then for short-range $V^{(ij)}$ coefficients of the differential operator $(V - V^a) M_a$ are also short-range (in the configuration space of the N-particle system). It is the only place where the short-range assumption on pair potentials is used. By property 2^0 the function $[V^a, M_a]$ is also short-range. Thus the operator $V M_a - M_a V^a$ can be taken into account by the limiting absorption principle. The commutator $[H_0, M_a]$ factorizes into a product of H_a- and H-smooth operators according to the radiation estimates.

Similar arguments show that, for $\sum_a m_a = m$ and $M = \sum_a M_a$ (the sums here are taken over all possible break-ups of the N-particle system), the wave operator (observable) $W^{\pm}(H, H; \pm M E(\Lambda))$ also exists. Moreover, it can be easily achieved that $m(x) \geq 1$. Then it follows from the Mourre estimate that $W^{\pm}(H, H; \pm M E(\Lambda))$ is positively definite on the subspace $E(\Lambda)\mathcal{H}$ and hence its range coincides with this subspace. It means that for any $f \in E(\Lambda)\mathcal{H}$

$$\lim_{t \to \pm\infty} || \exp(-iHt) f - M \exp(-iHt) g^{\pm} || = 0 \quad \text{if} \quad f = W^{\pm}(H, H; M_a E(\Lambda)) g^{\pm}.$$

The existence of the wave operators $W^{\pm}(H_a, H; M_a E(\Lambda))$ implies that for any $g^{\pm} = E(\Lambda) g^{\pm}$

$$\lim_{t \to \pm\infty} ||M \exp(-iHt) g^{\pm} - \sum_a \exp(-iH_a t) g_a^{\pm}|| = 0, \quad g_a^{\pm} = W^{\pm}(H_a, H; M_a E(\Lambda)) g^{\pm}.$$

Combining the results obtained, we see that $\exp(-iHt) f$ decomposes asymptotically into simpler evolutions $\exp(-iH_a t) g_a^{\pm}$. This is one of equivalent formulations of the asymptotic completeness.

5. Thus, strictly speaking, both long-range and multiparticle scattering problems surpass the perturbation theory framework. Both of them were open for a sufficiently long time, but were essentially solved during last twenty years. One of the goals of this survey is to present a unified approach to solutions of both these problems.

As discussed above, it is based on the limiting absorption principle and radiation estimates and fits these problems into the framework of smooth scattering theory.

Another goal is to describe the stationary approach where the evolution operators $\exp(-iH_0t)$ and $\exp(-iHt)$ are expressed via the resolvents of the operators H_0 and H. Then in place of the limits as $t \to \pm\infty$ defining the wave operators $W^{\pm}(H, H_0; J)$ one has to study boundary values of the resolvents as the spectral parameter approaches the continuous spectrum. Of course, such limits cannot exist in the space $\mathcal{H}$ but might exist in a proper couple of rigged spaces. To put it differently, $R(z)$ has boundary values when sandwiched by a suitable operator K. This means that the operator-function $KR(z)K^*$ is continuous in z up to the continuous spectrum of H. Practically, this condition is only slightly stronger than H-smoothness of the operator K. An important advantage of the stationary approach is that it allows us to obtain representations in terms of resolvents for different objects of the theory (wave and scattering operators, the scattering matrix). In particular, using these formulas we study the structure of the scattering matrix and its spectral properties.

Originally, this survey was intended to represent personal interests of the author and hence relied heavily on his own papers. However we have also reviewed different methods of scattering theory. Thus this survey is an instantaneous picture of the present state of scattering theory as seen by the author. A discussion of open problems is its important part. Naturally we took into account the existing literature (primarily v. 3 and 4 of [134]) and tried, as far as possible, to avoid overlap. We tried also to fill in the gaps in the literature (for example, we discuss scattering by unbounded obstacles). To a certain extent, this survey can be considered as complementary to the recent book [41] where the time-dependent method was adopted. In its present form the survey covers different branches of scattering theory but probably remains biased towards the author's interests. Its short version was published in [173].

The structure of the survey is clear from the contents. Moreover, each part contains its own short introduction.

The idea to write a survey based on the author's talk at ICM'98 is due to M. S. Birman. Some parts of the manuscipt were discussed with him. The author is deeply grateful to M. S. Birman.

Contents

Part III. The multiparticle Schrödinger operator and related problems 86

Part I

The Schrödinger operator of two-particle systems

We try to concentrate on parts of the theory of the two-particle Schrödinger operator not well represented in standard textbooks on this subject (see e.g. [6], [134]). It is relatively difficult in the short-range case already widely discussed in the literature. However, for example, the proof of the absence of the singular continuous spectrum in Section 2 is new, and Section 3 contains a lot of miscellaneous material.

On the other hand, asymptotic completeness in the long-range case is one of difficult problems of scattering theory solved only recently. The stationary approach to this problem is almost not discussed in the monographic literature (Chapter 30 of [67] seems to be the only exception). We consider two essentially different versions of the stationary approach to this problem. The first of them (Section 4) fits it in the general framework of smooth scattering theory. In addition to the limiting absorption principle and radiation estimates mentioned already in the Introduction, this requires the pseudo-differential operators calculus. Another version (Section 5) relies on a preliminary construction of eigenfunctions of the continuous spectrum of the Schrödinger operator.

For a discussion of time-dependent methods in two-particle scattering see [129, 41].

1. BASIC CONCEPTS

We give only a very brief presentation of basic concepts of scattering theory since they can be found, e.g., in the books [97, 134, 8, 165].

1.1. Classification of the spectrum. Scattering theory can be considered as a part of a more general perturbation theory. The ideology of the latter is that detailed information about an "unperturbed" self-adjoint operator H_0 enables us to draw conclusions about another self-adjoint operator H provided H_0 and H differ in an appropriate sense little from one another. Eigenvalues of H_0 can generically be shifted under arbitrary small perturbations, but the formulas for these shifts are basically the same as in the finite-dimensional case (in the linear algebra). The possible existence of the continuous spectrum is specific for infinite-dimensional spaces.

Scattering theory can be characterized as perturbation theory on this component of the spectrum and studies its remarkable stability. Very loosely speaking, some of the results of scattering theory can be considered as far advanced extensions of the H. Weyl theorem which states the stability of the essential spectrum under compact perturbations. Actually, the most interesting problems of scattering theory (for example, for the Schrödinger operator with long-range and multiparticle potentials) do not fit exactly into the framework of perturbation theory and lie on the border of its applicability.

Scattering theory requires classification of the spectrum based on measure theory. Recall that for any self-adjoint operator H with the spectral family $E = E_H$ in a Hilbert space $\mathcal{H}$, this space can be decomposed into the orthogonal sum

$$\mathcal{H} = \mathcal{H}^{(p)} \oplus \mathcal{H}^{(sc)} \oplus \mathcal{H}^{(ac)}$$

of invariant subspaces. The subspace $\mathcal{H}^{(p)}$ is spanned by eigenvectors of H and the subspaces $\mathcal{H}^{(sc)}$, $\mathcal{H}^{(ac)}$ are distinguished by the condition that the measure $(E(X)f, f)$ (here $X \subset \mathbb{R}$ is a Borel set) is singularly or absolutely continuous with respect to the Lebesgue measure for all $f \in \mathcal{H}^{(sc)}$ or $f \in \mathcal{H}^{(ac)}$. Typically (in applications to differential operators) the singularly continuous part is absent, that is $\mathcal{H}^{(sc)} = \{0\}$. We denote by $H^{(ac)}$ the restriction of H on its absolutely continuous subspace $\mathcal{H}^{(ac)} = \mathcal{H}_H^{(ac)}$ and by $P^{(ac)} = P_H^{(ac)}$ the orthogonal projection on this subspace.

By the spectral theorem, the resolvent $R(z) = (H - z)^{-1}$ of H is the Cauchy-Stieltjes integral

$$(R(z)f, f) = \int_{-\infty}^{\infty} (\lambda - z)^{-1} d(E(\lambda)f, f), \quad \operatorname{Im} z \neq 0, \quad \forall f \in \mathcal{H}. \tag{1.1}$$

It is convenient to introduce notation for its imaginary part

$$\delta_\varepsilon(H - \lambda) = (2\pi i)^{-1}(R(\lambda + i\varepsilon) - R(\lambda - i\varepsilon)).$$

The boundary values of function (1.1) exist and

$$\lim_{\varepsilon \to 0}(\delta_\varepsilon(H - \lambda)f, f) = \pi^{-1} \lim_{\varepsilon \to 0} \varepsilon ||R(\lambda \pm i\varepsilon)f||^2 = (dE(\lambda)f, f)/d\lambda \tag{1.2}$$

for almost all $\lambda \in \mathbb{R}$. Of course, the set of full measure, where both sides of (1.2) are correctly defined and equality (1.2) holds, depends on the vector f. We note also an expression for the operator $P^{(ac)}$:

$$(P^{(ac)}f, f) = \int_{-\infty}^{\infty} d(E(\lambda)f, f)/d\lambda \, d\lambda, \quad \forall f \in \mathcal{H}. \tag{1.3}$$

All these relations for quadratic forms can naturally be extended to sesquilinear forms.

1.2. Wave operators. Scattering theory is concerned with the structure of the absolutely continuous component and resolves two related problems. The first of them is the study of the behaviour for large times of the time-dependent Schrödinger equation

$$i\partial u/\partial t = Hu, \quad u(0) = f. \tag{1.4}$$

The second problem consists in finding conditions for the unitary equivalence of the absolutely continuous parts $H_0^{(ac)}$ and $H^{(ac)}$ of the operators H_0 and H. Of course, equation (1.4) has a unique solution $u(t) = \exp(-iHt)f$, while the solution of the same equation with the "free" operator H_0 is given by the formula $u_0(t) = \exp(-iH_0t)f_0$. From the viewpoint of scattering theory the function $u(t)$ has "free" asymptotics as $t \to \pm\infty$ if

$$\lim_{t\to\pm\infty} ||u(t) - u_0^{\pm}(t)|| = 0, \quad u_0^{\pm}(t) = \exp(-iH_0t)f_0^{\pm}, \tag{1.5}$$

for appropriate initial data $f_0^{\pm}$. Here and everywhere a relation containing the signs " $\pm$ " is understood as two independent equalities. Relation (1.5) leads to a connection between the corresponding initial data $f_0^{\pm}$ and f,

$$f = \lim_{t\to\pm\infty} \exp(iHt)\exp(-iH_0t)f_0^{\pm},$$

and motivates the following fundamental definition. At a formal level it was given by C. Møller [119] and was made precise by K. Friedrichs [52].

Definition 1.1 *The wave operator for a pair of self-adjoint operators H_0 and H is the operator*

$$W^{\pm} = W^{\pm}(H, H_0) = s - \lim_{t\to\pm\infty} \exp(iHt)\exp(-iH_0t)P_0, \quad P_0 := P_{H_0}^{(ac)}, \tag{1.6}$$

provided that the corresponding strong limit exists.

Obviously, it suffices to verify the existence of these limits on any set dense in the absolutely continuous subspace $\mathcal{H}_0^{(ac)}$ of the operator H_0. The wave operator is isometric on $\mathcal{H}_0^{(ac)}$ and enjoys the intertwining property $HW^{\pm} = W^{\pm}H_0$, so its range $\operatorname{Ran} W^{\pm}$ is contained in the absolutely continuous subspace $\mathcal{H}^{(ac)}$ of the operator H.

Definition 1.2 *The operator $W^{\pm}(H, H_0)$ is said to be complete if*

$$\operatorname{Ran} W^{\pm} = \mathcal{H}^{(ac)}. \tag{1.7}$$

It is easy to see that the completeness of $W^{\pm}(H, H_0)$ is equivalent to the existence of the "inverse" wave operator $W^{\pm}(H_0, H)$. Thus, if the wave operator $W^{\pm}(H, H_0)$ exists and is complete, then the operators $H_0^{(ac)}$ and $H^{(ac)}$ are unitarily equivalent. This shows that both problems of scattering theory are intimately related. We emphasize that scattering theory studies not arbitrary unitary equivalence but only the "canonical" one realized by the wave operators.

An important generalization of the definition above to the case of self-adjoint operators acting in different spaces was suggested by T. Kato [98].

Definition 1.3 *Let H_0 and H be self-adjoint operators on Hilbert spaces $\mathcal{H}_0$ and $\mathcal{H}$, respectively, and let "identification" $J : \mathcal{H}_0 \to \mathcal{H}$ be a bounded operator. Then the wave operator for the triple H_0, H and J is defined by the equality*

$$W^{\pm} = W^{\pm}(H, H_0; J) = s - \lim_{t\to\pm\infty} \exp(iHt) J \exp(-iH_0 t) P_0, \tag{1.8}$$

provided again that this strong limit exists.

Identifications can of course be different for $t \to \infty$ and $t \to -\infty$; in this case $W^{\pm} = W^{\pm}(H, H_0; J^{\pm})$. The intertwining property $HW^{\pm} = W^{\pm}H_0$ is preserved for wave operator (1.8), and it is isometric on $\mathcal{H}_0^{(ac)}$ if and only if

$$\lim_{t\to\pm\infty} ||J \exp(-iH_0 t) f_0|| = ||f_0|| \tag{1.9}$$

for any $f_0 \in \mathcal{H}_0^{(ac)}$. Since

$$s - \lim_{|t|\to\infty} K \exp(-iH_0 t) P_0 = 0 \tag{1.10}$$

for a compact operator K, the wave operators (1.8) corresponding to identifications J_1 and J_2 coincide if $J_2 - J_1$ is compact or, at least, $(J_2 - J_1)E_0(X)$ is compact for any bounded interval X. Consideration of wave operators (1.8) with $J \neq I$ may of course be of interest also in the case $\mathcal{H}_0 = \mathcal{H}$.

As before, the wave operator (1.8) is called complete if equality (1.7) holds, which imposes rather stringent conditions on J. In the general case we define J-completeness. Remark that the set $N^{\pm}$ of elements $f \in \mathcal{H}$ such that

$$\lim_{t\to\pm\infty} ||J^* \exp(-iHt) P f|| = 0, \quad P := P_H^{(ac)},$$

is a subspace, and $N^{\pm}$ is orthogonal to the subspace $\overline{\operatorname{Ran} W^{\pm}(H, H_0; J)}$.

Definition 1.4 *The wave operator $W^{\pm}(H, H_0; J)$ is called J-complete if*

$$N^{\pm} \oplus \overline{\operatorname{Ran} W^{\pm}(H, H_0; J)} = \mathcal{H}.$$

It is not difficult to check that J-completeness of $W^{\pm}(H, H_0; J)$ and the existence of the auxiliary wave operator $W^{\pm}(H_0, H_0; J^*J)$ are equivalent to the existence of the wave operator $W^{\pm}(H_0, H; J^*)$. Of course,

$$W^{\pm}(H, H_0; J)^* = W^{\pm}(H_0, H; J^*)$$

if both these wave operators exist.

In the particular cases $J = E_0(\Lambda)$ and $J = E(\Lambda)E_0(\Lambda)$ where $\Lambda \subset \mathbb{R}$ is some Borel set, the wave operators (1.8) appeared first in the papers of M. Sh. Birman [12], [13]. Clearly, the wave operator $W^{\pm}(H, H_0; E_0(\Lambda))$ is isometric on $E_0(\Lambda)P_0\mathcal{H}$ and

$$\operatorname{Ran} W^{\pm}(H, H_0; E_0(\Lambda)) \subset E(\Lambda)P\mathcal{H}.$$

Naturally, it is called complete if equality holds here. "Local" wave operators are also useful when considering "global" wave operators. For example, the existence of $W^{\pm}(H, H_0; JE_0(\Lambda_n))$ implies the existence of $W^{\pm}(H, H_0; J)$ provided the union of the sets Λ_n exhausts $\mathbb{R}$ up to a set of Lebesgue measure zero.

We note also the chain rule or multiplication theorem of wave operators.

Proposition 1.5 *If $W^{\pm}(H_1, H_0; J_0)$ and $W^{\pm}(H, H_1; J_1)$ exist, then the wave operator $W^{\pm}(H, H_0; J_1J_0)$ also exists and*

$$W^{\pm}(H, H_0; J_1J_0) = W^{\pm}(H, H_1; J_1)W^{\pm}(H_1, H_0; J_0). \tag{1.11}$$

The following simple but convenient condition [33] for the existence of wave operators (1.8) is usually called Cook's criterion.

Proposition 1.6 *Suppose that the operator J maps the domain $\mathcal{D}(H_0)$ of the operator H_0 into $\mathcal{D}(H)$. Let*

$$\int_0^{\pm\infty} ||(HJ - JH_0)\exp(-iH_0t)f||dt < \infty$$

for all f from some set $D_0 \subset \mathcal{D}(H_0) \cap \mathcal{H}_0^{(ac)}$ dense in $\mathcal{H}_0^{(ac)}$. Then the wave operator $W^{\pm}(H, H_0; J)$ exists.

Along with the wave operators an important role in scattering theory is played by the operator

$$\mathbf{S} = \mathbf{S}(H, H_0; J^+, J^-) = W^+(H, H_0; J^+)^* W^-(H, H_0; J^-), \tag{1.12}$$

known as the scattering operator. It commutes with H_0 and hence reduces to multiplication by the operator-function $S(\lambda) = S(\lambda; H, H_0; J^+, J^-)$, called the scattering matrix, in a representation of $\mathcal{H}_0$ which is diagonal for H_0. The scattering operator (1.12) is unitary on the subspace $\mathcal{H}_0^{(ac)}$ provided the wave operators $W^{\pm}(H, H_0; J^{\pm})$ exist, are isometric and complete. In the case $\mathcal{H}_0 = \mathcal{H}$, $J^{\pm} = I$ (the identity operator), the scattering operator $\mathbf{S}(H, H_0)$ connects the asymptotics of the solutions of equation (1.4) as $t \to -\infty$ and as $t \to +\infty$ in terms of the free problem, that is $\mathbf{S}(H, H_0) : f_0^- \mapsto f_0^+$ where $f_0^{\pm}$ are defined in (1.5). The scattering operator and the scattering matrix are usually of great interest in mathematical physics problems, because they connect the "initial" and the "final" characteristics of the process directly, bypassing its consideration for finite times. This also explains the term "scattering theory" which is borrowed from physics.

1.3. Eigenfunction expansions. The approach in scattering theory relying on definition (1.6) (or, more generally, (1.8)) is called time-dependent. An alternative possibility is to change the definition of wave operators replacing the unitary groups by the corresponding resolvents $R_0(z) = (H_0 - z)^{-1}$ and $R(z) = (H - z)^{-1}$. They are related by a simple identity

$$R(z) = R_0(z) - R_0(z)VR(z) = R_0(z) - R(z)VR_0(z), \quad \operatorname{Im} z \neq 0, \tag{1.13}$$

where $V = H - H_0$. In this, stationary, approach in place of the limits (1.6) one has to study the boundary values (in a suitable topology) of the resolvents as the spectral parameter z tends to the real axis. The stationary approach goes back to K. Friedrichs (see [52] and his book [53]). An important advantage of the stationary approach is that it gives convenient formulas for the wave operators and the scattering matrix. Note that rather an advanced scattering theory can be developed (see e.g. [21]) if the existence of necessary boundary values is taken as an a priori assumption.

Let us illustrate the stationary scheme on the example of the Schrödinger operator $H = -\Delta + V(x)$, $V = \bar{V}$, acting in the space $\mathcal{H} = L_2(\mathbb{R}^d)$. The results below are due essentially to A. Ya. Povzner (see [131, 132] and [69]). The temporal asymptotics of solutions of the time-dependent Schrödinger equation (1.4) is closely related to the asymptotics at large distances of solutions of the stationary Schrödinger equation

$$-\Delta\psi + V(x)\psi = \lambda\psi. \tag{1.14}$$

If

$$|V(x)| \leq C(1+|x|)^{-\rho} \tag{1.15}$$

where $\rho > d$, then for any $\lambda > 0$ and any unit vector $\omega \in \mathbb{S}^{d-1}$ equation (1.14) has a (unique) solution with the asymptotics

$$\psi(x;\omega,\lambda) = \exp(i\lambda^{1/2}\langle\omega,x\rangle) + a|x|^{-(d-1)/2}\exp(i\lambda^{1/2}|x|) + o(|x|^{-(d-1)/2}) \tag{1.16}$$

as $|x| \to \infty$. The coefficient $a = a(\hat{x},\omega;\lambda)$ depends on the incident direction ω of the incoming plane wave $\psi_0(x;\omega,\lambda) = \exp(i\lambda^{1/2}\langle\omega,x\rangle)$ (which satisfies of course the free equation $-\Delta\psi_0 = \lambda\psi_0$), its energy λ and the direction $\hat{x} = x|x|^{-1}$ of observation of the outgoing spherical wave $|x|^{-(d-1)/2}\exp(i\lambda^{1/2}|x|)$. The function $a(\theta,\omega;\lambda)$ is called the scattering amplitude. It is continuous in $\theta,\omega \in \mathbb{S}^{d-1}$ and $\lambda > 0$ and can be expressed via the solution $\psi(x;\omega,\lambda)$ by the formula

$$a(\theta,\omega;\lambda) = -e^{\pi i(d-3)/4}2^{-1}(2\pi)^{-(d-1)/2}\lambda^{(d-3)/4}\int_{\mathbb{R}^d}\exp(-i\lambda^{1/2}\langle\theta,x\rangle)V(x)\psi(x;\omega,\lambda)dx. \tag{1.17}$$

In terms of boundary values of the resolvent the scattering solution, or eigenfunction of the continuous spectrum, $\psi(x;\omega,\lambda) = \psi^-(x;\omega,\lambda)$ can be constructed by the formula

$$\psi^{\pm}(\omega,\lambda) = \psi_0(\omega,\lambda) - R(\lambda \mp i0)V\psi_0(\omega,\lambda). \tag{1.18}$$

Functions (1.18) obviously satisfy equation (1.14). Using the resolvent identity (1.13), it is easy to derive the Lippmann-Schwinger equation

$$\psi^{\pm}(\omega,\lambda) = \psi_0(\omega,\lambda) - R_0(\lambda \mp i0)V\psi^{\pm}(\omega,\lambda) \tag{1.19}$$

for $\psi^{\pm}(\omega,\lambda)$. Here $R_0(z)$ is the resolvent of the unperturbed operator $H_0 = -\Delta$. Hence the asymptotics (1.16) can be deduced from that of the free Green function (integral kernel of the operator $R_0(\lambda \pm i0)$). Actually, say, for any $f \in C_0^\infty(\mathbb{R}^d)$,

$$(R_0(\lambda \pm i0)f)(x) = c_{\pm}(\lambda)(\Gamma_0(\lambda)f)(\pm\hat{x})|x|^{-(d-1)/2}e^{\pm i\lambda^{1/2}|x|} + O(|x|^{-(d+1)/2}), \tag{1.20}$$

where $c_\pm(\lambda) = \pi^{1/2}\lambda^{-1/4}e^{\mp i\pi(d-3)/4}$ and

$$(\Gamma_0(\lambda)f)(\omega) = 2^{-1/2}\lambda^{(d-2)/4}\hat{f}(\lambda^{1/2}\omega), \quad \omega \in \mathbb{S}^{d-1}, \tag{1.21}$$

is (up to the numerical factor) the restriction of the Fourier transform $\hat{f}$ onto the sphere of radius $\lambda^{1/2}$. The asymptotics of the radial derivative $\psi_r^\pm(x;\omega,\lambda)$ of function (1.18) can be found quite similarly. We emphasize that the asymptotics at infinity of ψ and of ψ_r determine a unique solution of equation (1.14). Note also that

$$\psi^+(x;\omega,\lambda) = \overline{\psi^-(x;-\omega,\lambda)}.$$

The wave operators can be constructed in terms of solutions $\psi^\pm$. Set $\xi = \lambda^{1/2}\omega$ (ξ is the momentum variable), write $\psi^\pm(x,\xi)$ instead of $\psi^\pm(x;\omega,\lambda)$ and consider two transformations

$$(\mathcal{F}^\pm f)(\xi) = (2\pi)^{-d/2}\int_{\mathbb{R}^d}\overline{\psi^\pm(x,\xi)}f(x)dx \tag{1.22}$$

(defined initially, for example, on the Schwartz class $\mathcal{S}(\mathbb{R}^d)$) of the space $L_2(\mathbb{R}^d)$ into itself (we do not distinguish in notation the configuration and momentum spaces). The operators $\mathcal{F}^\pm$ can be regarded as generalized Fourier transforms, and both of them coincide with the usual Fourier transform $\mathcal{F}_0 = \mathcal{F}$ if $V = 0$. It follows from (1.14), (1.22) that under the action of $\mathcal{F}^\pm$ the operator H goes over into multiplication by $|\xi|^2$, i.e.,

$$\mathcal{F}^\pm H = |\xi|^2\mathcal{F}^\pm.$$

Moreover, $\mathcal{F}^\pm$ is an isometry on $\mathcal{H}^{(ac)}$, it is zero on $\mathcal{H}\ominus\mathcal{H}^{(ac)}$ and its range $\operatorname{Ran}\mathcal{F}^\pm = L_2(\mathbb{R}^d)$. It can also be shown that the vector

$$\left((\mathcal{F}^\pm)^* - \mathcal{F}_0^*\right)\exp(-i|\xi|^2t)\hat{f}$$

tends to zero as $t \to \pm\infty$ for arbitrary $\hat{f} = \mathcal{F}_0 f$. This implies the existence of the wave operators $W^\pm = W^\pm(H, H_0)$ for the pair H_0, H and gives the representation

$$W^\pm = (\mathcal{F}^\pm)^*\mathcal{F}_0. \tag{1.23}$$

Completeness of $W^\pm$ follows from this equality. Formula (1.23) is an example of a stationary representation for the wave operator. Formally, it means that

$$W^\pm\psi_0(\omega,\lambda) = \psi^\pm(\omega,\lambda).$$

The scattering matrix $S(\lambda)$ for the pair H_0, H can be computed in terms of the scattering amplitude. Namely, $S(\lambda)$ acts in the space $L_2(\mathbb{S}^{d-1})$, and $S(\lambda) - I$ is the integral operator whose kernel is the scattering amplitude. More precisely,

$$(S(\lambda)f)(\theta) = f(\theta) + ie^{\pi i(d-3)/4}\lambda^{(d-1)/4}(2\pi)^{-(d-1)/2}\int_{\mathbb{S}^{d-1}} a(\theta,\omega;\lambda)f(\omega)d\omega. \tag{1.24}$$

The scheme presented here can be directly generalized (see subsection 3.5) to potentials satisfying (1.15) for $\rho > (d+1)/2$, but it definitely fails for the general short-range case when $\rho > 1$ only. For an arbitrary $\rho > 1$, the wave operators $W^{\pm}$ exist and are complete (see subsection 2.1), but one cannot hope to construct solutions with asymptotics (1.16). In particular, formula (1.18) makes no sense if $\rho \leq (d+1)/2$.

Nevertheless, for any $\rho > 1$, the Schrödinger equation (1.14) has solutions corresponding to smearing of $\psi^{\pm}(\omega, \lambda)$ over ω. This allows one to obtain generalizations of transformations (1.22) and of representations (1.23) for the wave operators $W^{\pm}(H, H_0)$. This construction is discussed in subsection 2.3.

Let us, finally, make some general remarks about different forms of eigenfunction expansions. The spectral theorem states that every self-adjoint operator H has a "basis of eigenfunctions". This general assertion contains however no information neither on the spectrum nor on the eigenfunctions of the operator H. In applications to the Schrödinger operator, (1.22) gives probably the most advanced form of the expansion theorem. It establishes one-to-one correspondence between eigenfunctions of the operators H_0 and H, and (1.16) gives rather a detailed asymptotics of eigenfunctions of H as $|x| \to \infty$. Other possible forms of the expansion theorem are intermediary between these two levels. For example, the expansion theorem of subsection 2.3 is also intimately related to the wave operators, but it contains less information about eigenfunctions of the continuous spectrum.

1.4. The trace-class method. In scattering theory there are two essentially different approaches. One of them, the trace-class method, makes no assumptions about the "unperturbed" operator H_0. Its basic result is the following theorem of Kato and Rosenblum (see [92, 93] and [135]).

Theorem 1.7 *If $V = H - H_0$ belongs to the trace class $\mathfrak{S}_1$, then the wave operators $W^{\pm}(H, H_0)$ exist and are complete.*

We recall that the class $\mathfrak{S}_p$, $p \geq 1$, consists of compact operators T such that the norm

$$||T||_p = (\sum_k \lambda_k^p(|T|))^{1/p}, \quad |T| = (T^*T)^{1/2}, \tag{1.25}$$

is finite. Eigenvalues $\lambda_k(|T|) =: s_k(T)$ of a non-negative operator $|T|$ are called singular numbers of T. In particular, $\mathfrak{S}_1$ is the trace class and $\mathfrak{S}_2$ is the Hilbert-Schmidt class; the class of compact operators is denoted $\mathfrak{S}_\infty$.

According to the following Weyl-von Neumann-Kuroda theorem the condition $V \in \mathfrak{S}_1$ in Theorem 1.7 cannot be relaxed in the framework of operator ideals $\mathfrak{S}_p$.

Theorem 1.8 *Let H_0 be an arbitrary self-adjoint operator. For any $p > 1$ and any $\varepsilon > 0$ there exists a self-adjoint operator $V = V_{p,\varepsilon}$ such that $V \in \mathfrak{S}_p$, $||V||_p < \varepsilon$ and the operator $H = H_0 + V$ has purely point spectrum.*

This result demonstrates the sharpness of the condition $V \in \mathfrak{S}_1$ in Theorem 1.7. Indeed, under the assumptions of Theorem 1.8 the operator H has no absolutely continuous part. At the same time the operator H_0 may be absolutely continuous. Under these circumstances the wave operators $W^{\pm}(H, H_0)$ cannot exist.

Although sharp in the abstract framework, the Kato-Rosenblum theorem cannot be directly applied to the theory of differential operators where a perturbation is usually an operator of multiplication and hence is not even compact. Its generalizations applicable to this theory are due mainly to M. Sh. Birman, T. Kato himself and S. T. Kuroda. Let us formulate two results of this type.

Theorem 1.9 *Suppose that*

$$(H-z)^{-k}-(H_0-z)^{-k}\in\mathfrak{S}_1$$

for some $k=1,2,\ldots$ *and all* z *with* $\operatorname{Im} z\neq 0$. *Then the wave operators* $W^{\pm}(H,H_0)$ *exist and are complete.*

Theorem 1.10 *Assume that* $\mathcal{D}(H)=\mathcal{D}(H_0)$ *or that* $\mathcal{D}(|H|^{1/2})=\mathcal{D}(|H_0|^{1/2})$. *Suppose that*

$$E(\Lambda)(H-H_0)E_0(\Lambda)\in\mathfrak{S}_1$$

for any bounded interval Λ. *Then the wave operators* $W^{\pm}(H,H_0)$ *exist and are complete.*

Theorem 1.9 was proved in [16] for $k=1$ and in [95] for arbitrary k. Theorem 1.10 was obtained in [13].

We note also the following important result of M. Sh. Birman [11] known as the invariance principle for wave operators.

Theorem 1.11 *Suppose that* $\varphi(H)-\varphi(H_0)\in\mathfrak{S}_1$ *for a real function* φ *such that its derivative* φ' *is absolutely continuous and* $\varphi'(\lambda)>0$. *Then the wave operators* $W^{\pm}(H,H_0)$ *exist and*

$$W^{\pm}(H,H_0)=W^{\pm}(\varphi(H),\varphi(H_0)). \tag{1.26}$$

Note that the existence of $W^{\pm}(\varphi(H),\varphi(H_0))$ follows from Theorem 1.7. If $\varphi'(\lambda)<0$, then $W^{\pm}(H,H_0)$ in the left-hand side of (1.26) should be replaced by $W^{\mp}(H,H_0)$. Actually, it suffices to impose conditions on φ only on some set which, roughly speaking, coincides with the union of the spectra of the operators H_0 and H (the precise formulation may be found in [165]). The invariance principle can be extended to local wave operators $W^{\pm}(H,H_0;E_0(\Lambda))$ where Λ is an arbitrary interval. In this case the conditions on φ are imposed on Λ only.

A direct generalization of the Kato-Rosenblum theorem to the operators acting in different spaces is due to D. B. Pearson [128].

Theorem 1.12 *Suppose that* H_0 *and* H *are self-adjoint operators in spaces* $\mathcal{H}_0$ *and* $\mathcal{H}$, *respectively,* $J:\mathcal{H}_0\to\mathcal{H}$ *is a bounded operator and* $V=HJ-JH_0\in\mathfrak{S}_1$. *Then the wave operators* $W^{\pm}(H,H_0;J)$ *exist.*

Although rather sophisticated, the proof relies only on the following elementary lemma of M. Rosenblum.

Lemma 1.13 *For a self-adjoint operator H, consider the set $\mathfrak{R} \subset \mathcal{H}^{(ac)}$ of elements f such that*

$$r_H^2(f) := \operatorname{ess\,sup} d(E(\lambda)f, f)/d\lambda < \infty.$$

If $G : \mathcal{H} \to \mathcal{G}$ ($\mathcal{G}$ is some Hilbert space) is a Hilbert-Schmidt operator, then for any $f \in \mathfrak{R}$

$$\int_{-\infty}^{\infty} ||G \exp(-iHt)f||^2 dt \leq 2\pi r_H^2(f)||G||_2^2.$$

Moreover, the set $\mathfrak{R}$ is dense in $\mathcal{H}^{(ac)}$.

The Pearson theorem allows to simplify considerably the original proofs of different generalizations of the Kato-Rosenblum theorem (for example, of Theorems 1.9 and 1.10).

A consistent stationary approach was developed by M. Sh. Birman and S. B. Entina in the paper [15]. From analytical point of view it relies on the following result on boundary values of the resolvent which is important in its own sake.

Proposition 1.14 *Let H be a self-adjoint operator and let G_1, G_2 be arbitrary Hilbert-Schmidt operators. Then the operator-function $G_1 R(\lambda + i\varepsilon)G_2$ has a limit as $\varepsilon \to 0$ in the Hilbert-Schmidt class for almost all $\lambda \in \mathbf{R}$. Moreover, the operator-function $G_1 E(\lambda) G_2$ is differentiable in the trace norm for almost all $\lambda \in \mathbf{R}$.*

In particular, Proposition 1.14 allows us to obtain a stationary proof of the Pearson theorem (see [21] and the book [165]). This means the following. Remark that the existence of the strong wave operators $W^{\pm}(H, H_0; J)$ is equivalent to the existence of weak wave operators

$$\tilde{W}^{\pm}(H, H_0; J) = w - \lim_{t\to\pm\infty} P \exp(iHt) J \exp(-iH_0 t) P_0$$

(the limit is weak here), $\tilde{W}^{\pm}(H_0, H_0; J^*J)$ and to the equality

$$\tilde{W}^{\pm}(H, H_0; J)^* \tilde{W}^{\pm}(H, H_0; J) = \tilde{W}^{\pm}(H_0, H_0; J^*J). \tag{1.27}$$

The existence of all these weak wave operators is an immediate consequence of Lemma 1.13. The difficult part of the proof is a verification of equality (1.27), which is achieved by a stationary (involving the resolvents) computation.

We note that under the assumptions of Theorem 1.12 wave operators (1.8) are J-complete (see Definition 1.3) and the invariance principle holds

$$W^{\pm}(H, H_0; J) = W^{\pm}(\varphi(H), \varphi(H_0); J), \quad \varphi'(\lambda) > 0. \tag{1.28}$$

Remark also that the part of Proposition 1.14 concerning the spectral family $E(\lambda)$ is a basis for eigenfunction expansions of self-adjoint operators in an abstract setting [9].

The trace-class scattering theory was applied to differential operators in [106], [107], [14]. A typical result is given by the following

Theorem 1.15 *Suppose that*

$$\mathcal{H} = L_2(\mathbf{R}^d), \quad H_0 = -\Delta + V_0(x), \quad H = H_0 + V(x), \tag{1.29}$$

where the functions V_0 and V are real, $V_0 \in L_\infty(\mathbf{R}^d)$ and V satisfies estimate (1.15) *for some $\rho > d$. Then the wave operators $W^\pm(H, H_0)$ exist and are complete.*

Indeed, under assumptions of Theorem 1.15, conditions of both Theorems 1.9 and 1.10 can be verified (see e.g. [22]).

Another striking application of the trace-class theory concerns perturbations by higher order differential operators [14].

Theorem 1.16 *Let $\mathcal{H} = L_2(\mathbf{R}^d)$, $H_0 = -\Delta$ and $H = H_0 + V$, where*

$$V = \Delta(v(x)\Delta) = v(x)\Delta^2 + 2\langle \nabla v(x), \nabla \rangle \Delta + (\Delta v)(x)\Delta.$$

If the functions v, ∇v and Δv satisfy estimate (1.15) *for some $\rho > d$ and $v(x) \geq 0$, then the wave operators $W^\pm(H, H_0)$ exist and are complete.*

Note that the condition $v(x) \geq 0$ is necessary here, since otherwise a negative branch of the continuous spectrum may appear and the wave operators $W^\pm(H, H_0)$ (they still exist) are not complete. On the other hand, the conditions on ∇v and Δv can possibly be omitted in Theorem 1.16. The result of Theorem 1.16 seems to be inaccessible to smooth scattering theory discussed in the next subsection.

1.5. The smooth method. The second, smooth, method relies on a certain regularity of the perturbation in the spectral representation of the operator H_0. There are different ways to understand regularity. For example, in the Friedrichs-Faddeev model [49] H_0 acts as multiplication by independent variable in the space $\mathcal{H} = L_2(\Lambda; \mathfrak{N})$ where Λ is an interval (or the union of a finite number of disjoint intervals) and $\mathfrak{N}$ is an auxiliary Hilbert space. The perturbation V is an integral operator with sufficiently smooth kernel.

Another possibility is to use the concept of H-smoothness introduced by T. Kato in [96].

Definition 1.17 *An H- bounded operator K is called H-smooth if, for all $f \in \mathcal{D}(H)$,*

$$\int_{-\infty}^{\infty} ||K \exp(-iHt) f||^2 dt \leq C||f||^2.$$

It is important that this definition admits equivalent reformulations in terms of the resolvent or of the spectral family. Thus K is H-smooth if and only if

$$\sup_{\lambda \in \mathbf{R}, \varepsilon > 0} ||K(R(\lambda + i\varepsilon) - R(\lambda - i\varepsilon))K^*|| < \infty$$

or if and only if

$$\sup |X|^{-1} ||KE(X)||^2 < \infty \tag{1.30}$$

for any interval $X \subset \mathbf{R}$.

In applications the assumption of H-smoothness of an operator K imposes too stringent conditions on the operator H. In particular, the operator H is necessarily absolutely continuous if Ker$K$ (kernel of K) is trivial. This excludes eigenvalues and other singular points in the spectrum of H, for example, the bottom of the continuous spectrum for the Schrödinger operator with decaying potential or edges of bands if the spectrum has the band structure. The notion of local H-smoothness suggested by R. Lavine in [113] is considerably more flexible. By definition, K is called H-smooth on a Borel set $X \subset \mathbb{R}$ if the operator $KE(X)$ is H-smooth. Note that, under the assumption

$$\sup_{\lambda \in X, \varepsilon > 0} \|K(R(\lambda + i\varepsilon) - R(\lambda - i\varepsilon))K^*\| < \infty,$$

the operator K is H-smooth on the closure $\bar{X}$ of X.

The following assertion [96], [113] is simple but very useful.

Proposition 1.18 *Suppose that*

$$HJ - JH_0 = K^* K_0, \tag{1.31}$$

where the operators K_0 and K are H_0-smooth and H-smooth, respectively, on an arbitrary compact subinterval of some interval Λ. Then the wave operators

$$W^{\pm}(H, H_0; JE_0(\Lambda)) \quad \text{and} \quad W^{\pm}(H_0, H; J^* E(\Lambda))$$

exist (and are of course adjoint to each other).

If $HJ - JH_0 = \sum_{j=1}^{N} K_j^* K_{0,j}$ with H_0-smooth operators $K_{0,j}$ and H-smooth operators K_j, then the conclusion of Proposition 1.18 is preserved. This formally more general statement reduces to the former one if one introduces "vector" operators $K_0 = (K_{0,1}, \cdots, K_{0,N})$ and $K = (K_1, \cdots, K_N)$. Under assumptions of Proposition 1.18 the invariance principle, that is equality (1.28) with J replaced by $JE_0(\Lambda)$, also holds.

Of course Proposition 1.18 is not effective since the verification of H_0- and especially of H-smoothness may be a difficult problem. In simplest cases the following scheme is used. Suppose that the difference $V = H - H_0$ admits a factorization $V = K^* BK$ with a bounded operator $B = B^*$. Normally, H_0 -smoothness of the operator K can be checked directly. Then, under the assumption that $KR_0(z)K^* \in \mathfrak{S}_\infty$, $\operatorname{Im} z \neq 0$, one applies perturbative arguments to deduce its H-smoothness. Finally, Proposition 1.18 (for $J = I$ and $K_0 = BK$) guarantees the existence and completeness of the wave operators $W^{\pm}(H, H_0)$. This scheme is realized in subsection 2.1 on the example of the Schrödinger operator with a short-range potential.

Another important method to verify H-smoothness relies on consideration of the commutator $[H, M] = HM - MH$ of H with some auxiliary symmetric operator M. The crucial point is to find M such that the operator $i[H, M]$ is essentially positive. In this approach operators H_0 and H are treated at equal footing, so it does not fit into perturbation theory framework. There are at least two substantially different versions of the commutator method. The first of them is due mainly to C. R. Putnam [133] and T. Kato [99].

Proposition 1.19 *Suppose that*

$$G^*G \leq i[H, M] + K^*K, \tag{1.32}$$

where M is a bounded operator and K is H-smooth. Then G is also H-smooth.

Indeed, it follows from (1.32) that for any $f \in \mathcal{D}$ and any $T > 0$

$$\int_{-T}^{T} ||Ge^{-iHt}f||^2 dt \leq i \int_{-T}^{T} ([H, M]e^{-iHt}f, e^{-iHt}f)dt + \int_{-T}^{T} ||Ke^{-iHt}f||^2 dt. \tag{1.33}$$

The first term in the right-hand side equals

$$\int_{-T}^{T} d(Me^{-iHt}f, e^{-iHt}f)/dt\, dt = (Me^{-iHt}f, e^{-iHt}f)\Big|_{-T}^{T},$$

and hence is bounded by $2||M||\,||f||^2$. Since the second term is bounded by $C||f||^2$, the left-hand side of (1.33) admits the same estimate. Thus G satisfies Definition 1.17.

Proposition 1.19 is of interest already in the case $K = 0$. Moreover, it allows to find new H-smooth operators G given an H-smooth operator K. Proposition 1.19 can, in an obvious way, be "localized" on a bounded interval Λ. Indeed, multiplying (1.32) by $E(\Lambda)$ from the left and right and applying Proposition 1.19 to the operators $GE(\Lambda)$, $KE(\Lambda)$ and $E(\Lambda)ME(\Lambda)$ in place of G, K and M, respectively, we see that G is H-smooth on Λ provided M is H-bounded and K is H-smooth on Λ.

Another version of the commutator method where M is not H-bounded is due to E. Mourre [120]. It will be discussed in Sections 4 and 13 in applications to the Schrödinger operator with long-range and multiparticle potentials.

2. SHORT-RANGE INTERACTIONS. ASYMPTOTIC COMPLETENESS

Here we describe the smooth approach on the example of the Schrödinger operator $H = -\Delta + V$ in the space $\mathcal{H} = L_2(\mathbb{R}^d)$. Now V is multiplication by a real potential $V(x)$ satisfying (1.15) for some $\rho > 1$. Such potentials are called short-range. We first study the operator $H_0 = -\Delta$ which plays the role of an unperturbed or free operator. In the short-range case it is then possible to apply perturbative arguments to deduce a necessary information about the operator H from that for the operator H_0.

2.1. A perturbative scheme. Let $\langle x\rangle$ be multiplication by the function $(1+|x|^2)^{1/2}$ and let $L_2^{(l)} = L_2^{(l)}(\mathbb{R}^d)$ be the Hilbert space with the norm $||f||_{L_2^{(l)}} = ||\langle x\rangle^l f||_{L_2}$. Recall that the Sobolev space $\mathsf{H}^l = \mathsf{H}^l(\mathbb{R}^d)$ consists of functions $g(\xi)$ such that their inverse Fourier transforms $\check{g} = \mathcal{F}^* g \in L_2^{(l)}$ and $||g||_{\mathsf{H}^l} = ||\check{g}||_{L_2^{(l)}}$. Let the operator $\Gamma_0(\lambda)$ be defined by formula (1.21) and set $\mathfrak{N} = L_2(\mathbb{S}^{d-1})$. If $f \in L_2^{(l)}$ with $l > 1/2$, then, by the Sobolev trace theorem,

$$\left.\begin{aligned} ||\Gamma_0(\lambda)f||_{\mathfrak{N}} &\leq C||f||_{L_2^{(l)}}, \\ ||\Gamma_0(\lambda)f - \Gamma_0(\lambda')f||_{\mathfrak{N}} &\leq C|\lambda-\lambda'|^\alpha ||f||_{L_2^{(l)}}, \end{aligned}\right\} \tag{2.1}$$

for any $\alpha \leq l - 1/2$, $\alpha < 1$. Estimates (2.1) imply that the function

$$(E_0(\lambda)f, f) = \int_{|\xi|^2<\lambda} |\hat{f}(\xi)|^2 d\xi$$

is differentiable and

$$d(E_0(\lambda)f,f)/d\lambda = ||\Gamma_0(\lambda)f||^2_{\mathfrak{N}}, \quad f \in L_2^{(l)},\ l > 1/2, \tag{2.2}$$

is Hölder continuous in $\lambda > 0$ (uniformly in f, $||f||_{L_2^{(l)}} \leq 1$). Therefore applying the Privalov theorem to the Cauchy integral (1.1) for $(R_0(z)f,f)$ (it is of course taken over $[0,\infty)$), we obtain

Proposition 2.1 *The analytic operator-function*

$$\mathcal{R}_0(z) = \langle x\rangle^{-l} R_0(z) \langle x\rangle^{-l}, \quad l > 1/2, \tag{2.3}$$

considered in the space $\mathcal{H}$, is continuous in norm in the closed complex plane $\mathbb{C}$ cut along $[0,\infty)$ with possible exception of the point $z = 0$.

This assertion is called the limiting absorption principle for the operator H_0 and implies H_0 -smoothness of the operator $\langle x\rangle^{-l}$, $l > 1/2$, on any bounded disjoint from zero interval.

Proposition 2.2 *If Q is multiplication by a bounded function $q(x)$ tending to zero at infinity, then the operator $QR_0(z)$, $\operatorname{Im} z \neq 0$, is compact.*

This statement is equivalent to compactness of the integral operator K with kernel

$$k(x,\xi) = q(x)e^{i\langle x,\xi\rangle}(|\xi|^2 - z)^{-1}.$$

Let χ_r be multiplication by the characteristic function of the ball $|x| \leq r$ (or $|\xi| \leq r$). Clearly, the operator $K_r = \chi_r K \chi_r$ belongs to the Hilbert-Schmidt class and $||K_r - K|| \to 0$ as $r \to \infty$.

The limiting absorption principle for the operator H_0 ensures a similar but slightly weaker result for the operator H. To show it, we need the following assertion known as the analytic Fredholm alternative.

Proposition 2.3 *Let an operator-function $\mathbf{r}(z)$ acting in some Hilbert space be analytic in a half-band $\operatorname{Re} z \in (\alpha,\beta)$, $\operatorname{Im} z > 0$, and continuous in norm up to the line $\operatorname{Im} z = 0$. Suppose that $\mathbf{r}(z)$ is compact (or at least $\mathbf{r}^k(z)$ is compact for some $k = 2,3,\ldots$) and that -1 is not its eigenvalue. Then the set $\mathcal{N}$ of λ, where the homogeneous equation*

$$f + \mathbf{r}(\lambda + i0)f = 0$$

has a non-trivial solution, is closed in (α,β) and has the Lebesgue measure zero. The operator-function $(1 + \mathbf{r}(z))^{-1}$ is continuous in norm as z approaches the real axis at the points of $(\alpha,\beta) \setminus \mathcal{N}$.

The detailed proof of this result can be found e.g. in [165]. Its non-trivial part consists of proving that the Lebesgue measure of $\mathcal{N}$ is zero. If the operator-function $\mathbf{r}(z)$ were finite-dimensional, then we could apply a uniqueness theorem to the analytic function $\det(I + \mathbf{r}(z))$. Being continuous up to the boundary $\operatorname{Im} z = 0$ and not identically zero, it cannot vanish on a set of positive measure. In the general case we approximate the compact operator $\mathbf{r}(z)$ by finite-dimensional ones.

Let us return to the Schrödinger operator H. By the resolvent identity (1.13), the operator-function $\mathcal{R}(z) = \langle x\rangle^{-l}R(z)\langle x\rangle^{-l}$ considered in the space $\mathcal{H}$ satisfies the equation

$$\mathcal{R}(z) = (I + \mathcal{R}_0(z)B)^{-1}\mathcal{R}_0(z), \quad \operatorname{Im} z \neq 0, \tag{2.4}$$

where B is multiplication by the function $(1+|x|)^{2l}V(x)$. We choose $l = \rho/2$, so that B is a bounded operator. If

$$f + \mathcal{R}_0(z)Bf = 0,$$

then $\psi = R_0(z)\langle x\rangle^{-l}Bf$ satisfies the Schrödinger equation $H\psi = z\psi$. Since H is self-adjoint, this implies that $\psi = 0$ and hence $f = 0$. By Proposition 2.2, the operator $\mathcal{R}_0(z)$ is compact, so the inverse operator in (2.4) exists for all $\operatorname{Im} z \neq 0$. Thus the operator-function $(I + \mathcal{R}_0(z)B)^{-1}$ is analytic in the complex plane cut along $[0,\infty)$ with possible exception of poles on the negative half-axis. Positions of these poles coincide with eigenvalues of H. This implies that the negative spectrum of H consists of eigenvalues of finite multiplicity accumulating at the point 0 only.

Moreover, $(I+\mathcal{R}_0(z)B)^{-1}$ is continuous up to the cut $(0,\infty)$ except points $\lambda \in (0,\infty)$ where the homogeneous equation

$$f + \mathcal{R}_0(\lambda \pm i0)Bf = 0 \tag{2.5}$$

has a non-trivial solution. By Proposition 2.3, the set $\mathcal{N}_\pm$ of such points λ is closed in $(0,\infty)$ and has Lebesgue measure zero.

Put $\mathcal{N} = \mathcal{N}_+ \cup \mathcal{N}_-$ and $\Lambda = (0,\infty)\setminus\mathcal{N}$. The set Λ has full measure in $(0,\infty)$ and is open, so $\Lambda = \cup_n \Lambda_n$ where Λ_n are disjoint open intervals. It follows from equation (2.4) that the operator-function $\mathcal{R}(z)$ is continuous as z approaches the positive half-axis at the points of Λ. Therefore the operator $\langle x\rangle^{-l}$, $l > 1/2$, is H-smooth on any strictly interior subinterval of every Λ_n. Applying Proposition 1.18, we see that the wave operators $W^\pm(H, H_0; E_0(\Lambda_n))$ and $W^\pm(H_0, H; E(\Lambda_n))$ exist for all n. Since $E_0(\Lambda) = I$ and $E(\Lambda) = P$, this implies the existence of $W^\pm(H, H_0)$ and $W^\pm(H_0, H)$. Thus we obtain the following result.

Theorem 2.4 *Let* $H_0 = -\Delta$, $H = -\Delta + V$ *where* V *satisfies* (1.15) *with some* $\rho > 1$. *Then the wave operators* $W^\pm(H, H_0)$ *exist and are complete.*

The approach described briefly above can be considered as a refinement of arguments of A. Ya. Povzner [131, 132] and T. Ikebe [69]. It can be applied to a wide class of differential operators (see e.g. [110]). The proof of the completeness of $W^\pm(H, H_0)$ under optimal condition $\rho > 1$ in (1.15) was obtained by T. Kato [100]. He used the result of S. T. Kuroda on the Hölder continuity of function (2.2). First, it was derived in [108] by separating variables in spherical coordinates and studing kernels (they are expressed in terms of Bessel functions) of spectral families of the operators $-d^2/dr^2 + m(m+d-2)r^{-2}$ ($m = 0, 1, 2, \ldots$ is the orbital quantum number) in $L_2(\mathbb{R}_+)$. The idea to deduce this result from the Sobolev trace theorem appeared in [109]. Another analytical method (relying on a reduction to the one-dimensional case) of proving H_0-smoothness of the operator $\langle x\rangle^{-l}$, $l > 1/2$, was given by S. Agmon [1].

In general, only the behaviour of a potential at infinity is essential in scattering theory. Standard local singularities can easily be accommodated. Therefore we suppose that a potential is a bounded function. However, as shown by D. B. Pearson, very wild singularity of $V(x)$, say at $x = 0$, may lead to the break-down of completeness (see [127] or [134], v. 3).

2.2. The absence of the singular continuous spectrum. It follows from the continuity of $\mathcal{R}(z)$ that the operator H is absolutely continuous on the set $\Lambda = (0,\infty)\setminus\mathcal{N}$. Therefore the singular positive spectrum of H is necessarily contained in $\mathcal{N}$. Furthermore, one can prove

Theorem 2.5 *Under the assumption* (1.15) *where* $\rho > 1$, *the operator* $H = -\Delta + V$ *does not have the singular continuous spectrum. Positive eigenvalues of* H *have finite multiplicities and may accumulate at the point* 0 *only.*

For the proof of the first statement, it suffices to check that the set $\mathcal{N}$ consists of eigenvalues of the operator H. In terms of $u = \langle x\rangle^{-l}Bf$, $l = \rho/2$, equation (2.5) can be rewritten as

$$u + VR_0(\lambda \pm i0)u = 0. \tag{2.6}$$

Multiplying this equation by $R_0(\lambda \pm i0)u$, taking the imaginary part of the scalar product and using (1.2) for the operator H_0, we see that

$$\pi d(E_0(\lambda)u, u)/d\lambda = \operatorname{Im}(R_0(\lambda \pm i0)u, u) = 0.$$

Thus the Fourier transform $\hat{u}$ of u belongs to the Sobolev space H^l, $l > 1/2$, and

$$\hat{u}(\xi) = 0 \quad \text{for} \quad |\xi| = \lambda^{1/2}. \tag{2.7}$$

It follows from (2.6) that

$$\psi = R_0(\lambda \pm i0)u \tag{2.8}$$

is a formal (because of the singularity $(|\xi|^2 - \lambda \mp i0)^{-1}$) solution of the Schrödinger equation (1.14). Therefore one needs only to verify that $\psi \in L_2(\mathbb{R}^d)$. This is a direct consequence of (2.1) and (2.7) if $l > 1$. Such a choice of l is obviously possible if $\rho > 2$ and with some additional trick can be achieved for any $\rho > 3/2$. In the general case (see [25], for details) one uses two following analytical results.

Lemma 2.6 *Let Λ be a bounded disjoint from zero interval. Then for any $s \leq 1/2$ and any $\tau < (1-2s)^{-1}$*

$$\Big(\int_\Lambda ||h_\mu||_{\mathfrak{N}}^{2\tau} d\mu\Big)^{1/(2\tau)} \leq C||h||_{\mathsf{H}^s}, \quad h_\mu(\omega) = h(\mu^{1/2}\omega).$$

The proof follows from the complex interpolation between the cases $s = 0$ when $\tau = 1$ and $s > 1/2$ when $\tau = \infty$.

Lemma 2.7 *Let $g \in \mathsf{H}^l$ for $l \in (1/2, 1]$ and let $g(\xi) = 0$ for $|\xi| = \lambda^{1/2}$. Then the function $(|\xi|^2 - \lambda)^{-1}g(\xi)$ belongs to the space H^{-s} for any $s > 1 - l$.*

Indeed, by virtue of (2.1), for any $h \in \mathsf{H}^s$ we have that

$$\Big|\int_\Lambda (\mu - \lambda)^{-1}\Big(\int_{\mathfrak{N}} g(\mu^{1/2}\omega)\overline{h(\mu^{1/2}\omega)}d\omega\Big)d\mu\Big|$$
$$\leq \int_\Lambda |\mu - \lambda|^{-1}||g_\mu||_{\mathfrak{N}}||h_\mu||_{\mathfrak{N}}d\mu \leq C\int_\Lambda |\mu - \lambda|^{l-3/2}||h_\mu||_{\mathfrak{N}}d\mu \, ||g||_{\mathsf{H}^l}.$$

Using the Hölder inequality and applying Lemma 2.6, we estimate the last integral by $||h||_{\mathsf{H}^s}$ for any $s > 1 - l$. This gives the estimate

$$|(R_0(\lambda + i0)g, h)| \leq C\, ||g||_{\mathsf{H}^l}\, ||h||_{\mathsf{H}^s},$$

which is equivalent to Lemma 2.7.

Now one uses repeatedly equation (2.6) and Lemma 2.7 for $g = \hat{u}$. Starting from $\hat{u} \in \mathsf{H}^l$ where $l > 1/2$ only, we see that $\hat{u} \in \mathsf{H}^p$ for any $p < l + \rho - 1$. Thus, after

n steps, we obtain that $\hat{u} \in \mathsf{H}^p$ for any $p < l + n(\rho - 1)$. For n large enough, this implies that $\hat{u} \in \mathsf{H}^p$ for $p > 1$, and consequently function (2.8) belongs to $L_2(\mathbb{R}^d)$.

Similar arguments show that eigenvalues of H do not have positive accumulation points. For the proof of boundedness of the set of eigenvalues one uses additionally the estimate (see subsection 3.4)

$$||\mathcal{R}_0(\lambda \pm i0)|| = O(\lambda^{-1/2}), \quad \lambda \to \infty, \tag{2.9}$$

(see Corollary 3.6 where a stronger result is obtained). Finally, the multiplicity of an eigenvalue λ of H coincides with the dimension of all solutions of equation (2.5). Since the operator $\mathcal{R}_0(\lambda \pm i0)$ is compact, this number is finite.

Theorem 2.5 first appeared (under optimal condition $\rho > 1$ in (1.15)) in the paper by S. Agmon [1]. Even earlier, T. Kato [94] showed that under the assumptions of Theorem 2.4 the operator H does not have positive eigenvalues. In contrast to Theorems 2.4 and 2.5, this fact is specific for the Schrödinger operator. Let us collect together all these results.

Theorem 2.8 *Under the assumptions of Theorem 2.5, the positive spectrum of the operator H is absolutely continuous. The operator-function $\mathcal{R}(z) = \langle x\rangle^{-l} R(z) \langle x\rangle^{-l}$, $l > 1/2$, is continuous in norm in the closed complex plane $\mathbb{C}$ cut along $[0, \infty)$ with exception of the point $z = 0$ and negative eigenvalues of H.*

This theorem is known as the limiting absorption principle for the Schrödinger operator H.

2.3. Stationary representations of wave operators. In the general case $\rho > 1$ the Schrödinger equation (1.14) does not have solutions admitting asymptotic decomposition (1.16) into the sum of plane and outgoing spherical waves. In particular, representation (1.18) for scattering solutions $\psi(x; \omega, \lambda)$ definitely fails. Nevertheless one can construct solutions

$$u(x; \lambda) = \int_{\mathbb{S}^{d-1}} \psi(x; \omega, \lambda) a(\omega) d\lambda \tag{2.10}$$

of equation (1.14) which formally correspond to smearing of $\psi(\omega, \lambda)$ over $\omega \in \mathbb{S}^{d-1}$ with some function $a \in \mathfrak{N} = L_2(\mathbb{S}^{d-1})$.

Let us start with the free case and set

$$(\Gamma_0^*(\lambda) a)(x) = 2^{-1/2} \lambda^{(d-2)/4} (2\pi)^{-d/2} \int_{\mathbb{S}^{d-1}} \exp(i\lambda^{1/2} \langle x, \omega\rangle) a(\omega) d\omega. \tag{2.11}$$

For an arbitrary a (it may be a smooth function or even a distribution on the unit sphere), function (2.11) satisfies the equation $-\Delta u = \lambda u$, but its behaviour at infinity depends on regularity of a. For example, if a is a smooth function, then, by the stationary phase method, it it easy to check that function (2.11) decomposes asymptotically into sums of incoming and outgoing spherical waves. On the other hand, we recover the plane wave $\exp(i\lambda^{1/2} \langle \omega, x\rangle)$ if a is the Dirac-function. The operator $\Gamma_0^*(\lambda)$ is adjoint to operator (1.21). Therefore it follows from (2.1) that

$\Gamma_0^*(\lambda) : \mathfrak{N} \to L_2^{(-l)}(\mathbb{R}^d)$, $l > 1/2$, is a bounded operator and hence $\Gamma_0^*(\lambda)a \in L_2^{(-l)}(\mathbb{R}^d)$ for any $a \in \mathfrak{N}$.

The operator $H_0 = -\Delta$ can of course be diagonalized by the classical Fourier transform. To put it slightly differently, set

$$(F_0 f)(\lambda) = \Gamma_0(\lambda) f. \tag{2.12}$$

Then $F_0 : L_2(\mathbb{R}^d) \to L_2(\mathbb{R}_+; \mathfrak{N})$ is a unitary operator and $F_0 H_0 = \lambda F_0$.

Our goal here is to generalize this construction to the Schrödinger operator $H = -\Delta + V$. Put

$$\Gamma^\pm(\lambda) = \Gamma_0(\lambda)(I - VR(\lambda \pm i0)). \tag{2.13}$$

According to Theorem 2.8, it is correctly defined for $\lambda > 0$ as a bounded operator from $L_2^{(l)}(\mathbb{R}^d)$, $l > 1/2$, to $\mathfrak{N}$. Then we introduce the mapping (known as the generalized Fourier transform) by the formula

$$(F^\pm f)(\lambda) = \Gamma^\pm(\lambda) f. \tag{2.14}$$

Using (1.2), (1.13) and (2.2), it is easy to show that

$$\begin{aligned} ||\Gamma^\pm(\lambda) f||^2 &= \pi^{-1} \lim_{\varepsilon \to 0} \varepsilon ||R_0(\lambda \pm i\varepsilon)(I - VR(\lambda \pm i\varepsilon) f||^2 \\ &= \pi^{-1} \lim_{\varepsilon \to 0} \varepsilon ||R(\lambda \pm i\varepsilon) f||^2 = d(E(\lambda) f, f)/d\lambda. \end{aligned}$$

Integrating this equation over $\lambda \in [0, \infty)$ and taking into account (1.3), we see that

$$||F^\pm f||^2 = \int_0^\infty ||\Gamma^\pm(\lambda) f||_{\mathfrak{N}}^2 \, d\lambda = \int_0^\infty \frac{d(E(\lambda) f, f)}{d\lambda} d\lambda = ||Pf||^2, \quad P = E(0, \infty). \tag{2.15}$$

In particular, $F^\pm$ defined originally on the dense set $L_2^{(l)}(\mathbb{R}^d)$, $l > 1/2$, extends by continuity to a bounded operator $F^\pm : L_2(\mathbb{R}^d) \to L^2(\mathbb{R}_+; \mathfrak{N})$. Moreover,

$$(F^\pm)^* F^\pm = P. \tag{2.16}$$

The operators $F^\pm$ diagonalize the operator H, that is

$$F^\pm H f = \lambda F^\pm f, \tag{2.17}$$

because the functions (cf. (1.18))

$$\Gamma^\pm(\lambda)^* a = (I - R(\lambda \mp i0)V)\Gamma_0(\lambda)^* a \tag{2.18}$$

satisfy the Schrödinger equation (1.14). Applying the operator $I + R_0(\lambda \mp i0)V$ to both sides of (2.18), we see that equality $\Gamma^\pm(\lambda)^* a = 0$ ensures $\Gamma_0(\lambda)^* a = 0$ and hence $a = 0$. Thus $\operatorname{Ker} \Gamma^\pm(\lambda)^* = \{0\}$ for each $\lambda > 0$ which implies that $\operatorname{Ker}(F^\pm)^* = \{0\}$ and $\operatorname{Ran} F^\pm = L_2(\mathbb{R}_+; \mathfrak{N})$ or, equivalently, that

$$F^\pm (F^\pm)^* = I. \tag{2.19}$$

Finally, using the Parseval equality, we obtain that, for any elements f_0, f,

$$2\varepsilon \int_0^\infty e^{-2\varepsilon t}(\exp(\pm itH_0)f_0, \exp(\pm itH)f)dt = \pi^{-1}\varepsilon \int_{-\infty}^\infty (R_0(\lambda \pm i\varepsilon)f_0, R(\lambda \pm i\varepsilon)f)d\lambda, \tag{2.20}$$

where, in view of the resolvent identity (1.13),

$$\pi^{-1}\varepsilon(R_0(\lambda \pm i\varepsilon)f_0, R(\lambda \pm i\varepsilon)f) = (\delta_\varepsilon(H_0 - \lambda)f_0, (I - VR(\lambda \pm i\varepsilon))f). \tag{2.21}$$

Suppose now that $f_0, f \in L_2^{(l)}$ for some $l > 1/2$. Then, taking into account relations (1.2) (for the operator H_0) and (2.2), we can pass in (2.20) to the limit $\varepsilon \to 0$, which yields

$$(W^\pm f_0, f) = \int_0^\infty (\Gamma_0(\lambda)f_0, \Gamma^\pm(\lambda)f)d\lambda. \tag{2.22}$$

This is the stationary representation for the time-dependent wave operator $W^\pm$. Note that (2.22) can be rewritten (cf. (1.23)) as

$$W^\pm = (F^\pm)^* F_0. \tag{2.23}$$

Let us give a precise formulation of the results obtained.

Theorem 2.9 *Let assumptions of Theorem 2.4 be satisfied, and let $l > 1/2$. Then the operator $\Gamma^\pm(\lambda) : L_2^{(l)}(\mathbf{R}^d) \to \mathfrak{N}$ defined by formula (2.13) is bounded and depends continuously in norm on $\lambda > 0$. The operator $F^\pm : L_2(\mathbf{R}^d) \to L_2(\mathbf{R}_+; \mathfrak{N})$ defined by formula (2.14) on $L_2^{(l)}(\mathbf{R}^d)$ extends by continuity to a bounded operator, and it satisfies equalities (2.16), (2.17) and (2.19). The wave operators $W^\pm$ are related to $F^\pm$ by formula (2.22) (or (2.23)).*

Compared to the expansion theorem discussed in subsection 1.3, the present formulation has a weaker form. First, eigenfunctions (2.18) correspond to smearing of $\psi(x;\omega,\lambda)$ over ω, that is $(\Gamma^\pm(\lambda)^* a)(x)$ equals (2.10) (up to a numerical factor) if $\rho > d$. Second, in contrast to subsection 1.3 where $\psi(x;\omega,\lambda)$ was determined by its asymptotics as $|x| \to \infty$, we did not study here the asymptotics of functions (2.18). This drawback will be remedied in subsection 3.5. We emphasize nevertheless that the construction of this subsection (its details can be found in [165]), in particular a proof of (2.22), goes through avoiding the study of these asymptotics.

The stationary representation of wave operators given by formulas (2.13), (2.22) is actually of abstract nature. Its generalization to wave operators (1.8) may be found in subsection 7.2. On the other hand, for the Schrödinger operator, both with short- and long-range potentials, the operator $\Gamma^\pm(\lambda)$ can be defined in a different way (see subsection 5.1).

2.4. The Schrödinger operator on the lattice. The discrete analogue of the Schrödinger operator is completely similar to the continuous one. Actually, in view of applications to the Heisenberg model (see Section 16), we consider a more general class of discrete operators. Let now $\mathcal{H} = l_2(\mathbf{Z}^d)$ be the l_2-space of sequences $f(x)$, $x \in \mathbf{Z}^d$, on the lattice $\mathbf{Z}^d$ with the basis $e_1, \ldots, e_d$. We denote by $\mathrm{T}(p)$ the shift by a vector $p \in \mathbf{Z}^d$:

$$(\mathrm{T}(p)f)(x) = f(x+p).$$

The "free" operator is defined by the formula

$$H_0 = \sum_{j=1}^{d} \Big(\mu_j \mathrm{T}(e_j) + \bar{\mu}_j \mathrm{T}(-e_j) \Big), \tag{2.24}$$

where $\mu_j = |\mu_j| e^{i\varphi_j}$ are given complex numbers. In particular, H_0 reduces to the discrete Schrödinger operator if $\mu_j = 1$ for all $j = 1, \dots, d$. A perturbation V is introduced by the equality

$$V = \sum_{j=1}^{N} \Big(V_j \mathrm{T}(p_j) + \mathrm{T}(-p_j) V_j^* \Big), \tag{2.25}$$

where V_j are multiplications by functions $v_j(x)$ and p_j are given vectors. The operator V is in general non-local, but we obtain the operator of multiplication by a function $v(x)$ ($v = 2 \operatorname{Re} v_1$) if $N = 1$ and $p_1 = 0$. In contrast to the continuous case, the operator V is compact if $v_j(x) \to 0$ as $|x| \to \infty$, $j = 1, \dots, N$.

The operator H_0 can of course be diagonalized by the discrete Fourier transform:

$$(\mathcal{F} f)(\xi) = (2\pi)^{-d/2} \sum_{x \in \mathbb{Z}^d} f(x) e^{-i\langle x, \xi \rangle}, \quad \xi \in \mathbb{T}^d = [0, 2\pi)^d. \tag{2.26}$$

The mapping $\mathcal{F} : l_2(\mathbb{Z}^d) \to L_2(\mathbb{T}^d)$ is unitary, and the operator $\mathcal{F} H_0 \mathcal{F}^*$ acts as multiplication by the function

$$\hat{h}(\xi) = 2 \sum_{j=1}^{d} |\mu_j| \cos(\xi_j + \varphi_j). \tag{2.27}$$

Thus, the operator H_0 is bounded, and its spectrum coincides with the interval $[-\lambda_0, \lambda_0]$ where $\lambda_0 = 2 \sum_{j=1}^{d} |\mu_j|$. The critical points ξ, where $\nabla \hat{h}(\xi) = 0$, of function (2.27) have the coordinates $\xi_j = -\varphi_j \pmod{\pi}$. Hence its critical values (called thresholds) are points $2 \sum_{j=1}^{d} \tau_j |\mu_j|$, where $\tau_j = \pm 1$. Let us denote them $\lambda_1, \lambda_2, \dots, \lambda_p$, so that $\lambda_1 = -\lambda_0, \lambda_p = \lambda_0$.

The following result plays the role of Theorem 2.4 and can be proven in the same way.

Theorem 2.10 *Let H_0, V be given by formulas* (2.24), (2.25) *and $H = H_0 + V$. Suppose that all functions v_j satisfy the condition*

$$v_j(x) = O(|x|^{-\rho}), \quad \rho > 1, \quad x \in \mathbb{Z}^d. \tag{2.28}$$

Then the wave operators $W^{\pm}(H, H_0)$ exist and are complete.

Indeed, let $\langle x \rangle$ be multiplication by the function $(1 + |x|^2)^{1/2}$, $x \in \mathbb{Z}$. Then $V = \langle x \rangle^{-l} B \langle x \rangle^{-l}$, where $l = \rho/2 > 1/2$ and B is a bounded operator. Let us check (cf. Proposition 2.1) that the analytic operator-function $\langle x \rangle^{-l} R_0(z) \langle x \rangle^{-l}$ is continuous in norm in the closed complex plane $\mathbb{C}$ cut along $[-\lambda_0, \lambda_0]$ with possible exception of the points λ_k. Set $G_\lambda = \{\xi \in \mathbb{T}^d : \hat{h}(\xi) = \lambda\}$ and let ds_λ be the

Euclidian measure on the surface G_λ. By the Privalov theorem, it suffices to verify that the derivative (cf. (2.2))

$$(dE_0(\lambda)\langle x\rangle^{-l}f, \langle x\rangle^{-l}f)/d\lambda = \int_{G_\lambda} |g(\xi)|^2 |\nabla\hat{h}(\xi)|^{-1} ds_\lambda(\xi), \quad g = \mathcal{F}\langle x\rangle^{-l}f,$$

exists and is a Hölder continuous (uniformly in f, $||f|| \leq 1$,) function of $\lambda \in (\lambda_k, \lambda_{k+1})$ for $k = 1, 2, \dots, p-1$. For $g \in \mathsf{H}^l(\mathbb{T}^d)$, these statements follow again from the Sobolev trace theorem.

Since the operator V is compact, Proposition 2.3 is applicable, and the operator-function $\langle x\rangle^{-l}R(z)\langle x\rangle^{-l}$ is norm continuous up to the interval $[-\lambda_0, \lambda_0]$ except, possibly, a closed set of measure zero. Thus we can again refer to Proposition 1.18.

Also quite similarly to the continuous case, we can check an analogue of Theorem 2.5.

Theorem 2.11 *Under the assumptions of Theorem* 2.10, *the operator H does not have the singular continuous spectrum. Its eigenvalues different from the thresholds λ_k have finite multiplicities and may accumulate at the points λ_k only.*

Another proof of Theorems 2.10 and 2.11 relies on the Mourre estimate and may be found in [61].

2.5. An open problem. The assumptions of trace-class and smooth scattering theory are quite different. Thus it would be desirable to develop a theory unifying the trace-class and smooth approaches. Of course this problem admits different interpretations, but it becomes unambiguously posed in the context of applications, especially to differential operators. Consider, for example, the pair of operators (1.29). Smooth theory goes through if operator-function (2.3) is continuous in norm in the closed complex upper (and lower) half-plane with exception of some closed set $\mathcal{N} \subset \mathbb{R}$ of measure zero. Practically this requires an explicit spectral analysis of the operator H_0, which is possible for special V_0 only. The simplest case is of course $V_0 = 0$ but one can also accomodate long-range (see Section 10), multiparticle and periodic V_0 (see [25], where however rather burdensome and implicit conditions on V_0 were imposed, and [58]) potentials. Then the existence and completeness of the wave operators $W^\pm(H, H_0)$ is verified under assumption (1.15) on the perturbation V where $\rho > 1$. On the other hand, by Theorem 1.15, V_0 can be an arbitrary bounded function if $\rho > d$. Thus assumptions of smooth theory on an unperturbed operator H_0 are more stringent and on a perturbation V are less stringent then those of trace-class theory. This raises

Problem 2.12 *Let H_0 and H be given by* (1.29) *and let $d > 1$. Do the wave operators $W^\pm(H, H_0)$ exist for arbitrary $V_0 \in L_\infty(\mathbb{R}^d)$ and V satisfying bound* (1.15), *assuming only that $\rho > 1$?*

In the event of a positive solution of Problem 2.12, wave operators would be automatically complete under its assumptions. We conjecture, on the contrary, that Problem 2.12 has a negative solution. Moreover, we expect that the absolutely continuous part of the spectrum is no longer stable in the situation under consideration.

Problem 2.12 is linked to behaviour of solutions $u(t) = \exp(-iH_0 t)f$ of the time-dependent Schrödinger equation for large $|x|$ and $|t|$. If $H_0 = -\Delta$ (as well as for special choices of V_0 mentioned above), then the function $u(x,t)$ "lives" in the region where $|x| \geq c(f)|t|$ provided f is chosen from a suitable dense set. This means, for example, that

$$\int_{|x|\leq c(f)|t|} |u(x,t)|^2 dx$$

tends to zero as $|t| \to \infty$ (actually, faster than any power of $|t|^{-1}$). If $H_0 = -\Delta + V_0$ where V_0 is an arbitrary bounded function, then we can assert only that

$$\int_{-\infty}^{\infty} \Big(\int_{\mathbf{R}^d} (1+|x|)^{-2p} |u(x,t)|^2 dx \Big) dt < \infty, \quad 2p > d. \tag{2.29}$$

To check (2.29), we remark that for any bounded interval Λ the operator

$$(1+|x|)^{-p} E(\Lambda) \in \mathfrak{S}_2. \tag{2.30}$$

Indeed, if $d \leq 3$, then (2.30) follows from the obvious inclusion $(1+|x|)^{-p}(-\Delta + I)^{-1} \in \mathfrak{S}_2$. In the general case its proof can be found in [22]. Hence (2.29) for initial data $f \in \mathfrak{R} \cap E(\Lambda)\mathcal{H}$ is a consequence of Lemma 1.13. Roughly speaking, (2.29) means that $u(x,t)$ "lives" in the region where $|x| \geq c|t|^{1/d}$. Probably this result cannot be improved if V_0 is sufficiently wild.

If $u(x,t)$ is concentrated in the region where $|x| \sim |t|^s$, then $||Vu(t)|| \sim |t|^{-s\rho}$ and hence the wave operators $W^{\pm}(H, H_0)$ exist if $s\rho > 1$. Therefore the condition $\rho > d$ for the existence of these wave operators for an arbitrary bounded V_0 looks sufficiently plausible.

3. SHORT-RANGE INTERACTIONS. MISCELLANEOUS

3.1. The zero-energy resonance. The spectral point $z = 0$ was exluded from formulation of Theorem 2.8. Actually, the behaviour of the operator-function $\mathcal{R}(z) = \langle x\rangle^{-l}R(z)\langle x\rangle^{-l} : \mathcal{H} \to \mathcal{H}$ at this point is very sensitive to variations of the potential V of the Schrödinger operator $H = -\Delta + V$. In the free case $V = 0$, kernel of $R_0(z)$ is singular at $z = 0$ if $d = 1, 2$. On the contrary, if $d \geq 3$ and $l > 1$, then the operator $\mathcal{R}_0(z) = \langle x\rangle^{-l}R_0(z)\langle x\rangle^{-l} : \mathcal{H} \to \mathcal{H}$ is continuous in norm at this point. Of course, continuity of $\mathcal{R}(z)$ at $z = 0$ is incompatible with accumulation of negative eigenvalues of H at this point. This might happen if $\rho \leq 2$ in (1.15) .

The following assertion can be easily deduced from equation (2.4).

Proposition 3.1 *Let $d \geq 3$ and let (1.15) be satisfied with some $\rho > 2$. Assume that the homogeneous equation*

$$f + \mathcal{R}_0(0)Bf = 0 \tag{3.1}$$

has only a trivial solution $f = 0$. Then the operator-function $\mathcal{R}(z) = \langle x\rangle^{-l}R(z)\langle x\rangle^{-l}$, $l > 1$, is continuous in norm at the cut $[0, \infty)$.

If equation (3.1) has a non-trivial solution $f \in L_2(\mathbb{R}^d)$, then the Schrödinger equation $-\Delta\psi + V\psi = 0$ has a solution $\psi(x) = \langle x\rangle^l f(x)$ decaying as $|x| \to \infty$ but not necessarily from $L_2(\mathbb{R}^d)$. If $\psi \in L_2(\mathbb{R}^d)$, then of course $\lambda = 0$ is an eigenvalue of the operator H. In the opposite case we say that $\lambda = 0$ is a zero-energy resonance. Zero-energy resonances may exist only in dimensions $d \leq 4$. In the presence of a zero-energy resonance and $d = 3$, the kernel of the resolvent $R(z)$ has (see [154, 89, 124, 125]) the singularity

$$\psi(x)\overline{\psi(x')}(-z)^{-1/2}$$

as $z \to 0$. Here ψ satisfies the equation $-\Delta\psi + V\psi = 0$ and $\psi(x) \sim 2^{-1}\pi^{-1/2}|x|^{-1}$ as $|x| \to \infty$; the last condition is equivalent to the normalization

$$\int_{\mathbb{R}^3} V(x)\psi(x)dx = -2\pi^{1/2}.$$

3.2. General conditions for the existence of wave operators. For the existence of the wave operators $W^\pm(H, H_0)$ assumption (1.15) with $\rho > 1$ is not really necessary. Some part of $\mathbb{R}^d$ can be neglected. Moreover, a potential may be arbitrary singular, for example, on some plane.

Theorem 3.2 *Suppose that, for some closed cone N of measure zero, estimate (1.15) with $\rho > 1$ holds for all x from any closed cone K such that $K \cap N = \{0\}$. Let H be some self-adjoint operator such that $(Hg)(x) = -(\Delta g)(x) + V(x)g(x)$ for any $g \in C_0^\infty(\mathbb{R}^d \setminus N)$. Then the wave operators $W^\pm(H, H_0)$ exist.*

One of possible proofs of this assertion relies on the following remark. Recall that

$$(\exp(-iH_0t)f)(x) = (4\pi it)^{-d/2}\int_{\mathbf{R}^d}\exp(i(4t)^{-1}|x-x'|^2)f(x')dx' \qquad (3.2)$$

and set

$$(U_0(t)f)(x) = \exp(i(4t)^{-1}|x|^2)(2it)^{-d/2}\hat{f}(x/(2t)), \quad \hat{f} = \mathcal{F}f. \qquad (3.3)$$

Comparing (3.2) and (3.3), we see that

$$\lim_{t\to\pm\infty} ||\exp(-iH_0t)f - U_0(t)f|| = 0, \qquad (3.4)$$

so that the wave operators $W^{\pm}(H, H_0)$ and

$$\mathcal{W}^{\pm} = \mathcal{W}^{\pm}(H, H_0) = s - \lim_{t\to\pm\infty}\exp(iHt)U_0(t) \qquad (3.5)$$

exist simultaneously and $W^{\pm}(H, H_0) = \mathcal{W}^{\pm}(H, H_0)$. A direct calculation shows that, for $\hat{f} \in C_0^\infty(\mathbf{R}^d \setminus (\pm N))$,

$$((i\partial/\partial t - H)U_0(t)f)(x) = \exp(i(4t)^{-1}|x|^2)(2it)^{-d/2}$$
$$\times\big(-V(x)\hat{f}(x/(2t)) + (2t)^{-2}(\Delta\hat{f})(x/(2t))\big).$$

Thus the function $U_0(t)f$ is an approximate solution of the Schrödinger equation for $t \to \pm\infty$ in the sense that

$$\Big|\int_1^{\pm\infty} ||(i\partial/\partial t - H)U_0(t)f||dt\Big| < \infty. \qquad (3.6)$$

Since the sets $C_0^\infty(\mathbf{R}^d \setminus (\pm N))$ are dense in $L_2(\mathbf{R}^d)$, this implies (cf. Proposition 1.6) that limits (3.5) exist.

3.3. Anysotropically decaying potentials. Let us now consider a class of two-body potentials V with anysotropic fall-off at infinity. Suppose that

$$\mathbf{R}^d = X_1 \oplus X^1, \quad \dim X_1 = d_1,\ \dim X^1 = d^1,\ d_1 + d^1 = d, \quad x = (x_1, x^1), \qquad (3.7)$$

and

$$|V(x)| \le C(1+|x_1|)^{-\rho_1}(1+|x^1|)^{-\rho^1}, \quad \rho_1 > 0,\ \rho^1 > 0. \qquad (3.8)$$

The following result was obtained in [35].

Theorem 3.3 *Assume that* V *satisfies* (3.8) *where*

$$\left.\begin{array}{rcl} \rho_1 + 2^{-1}\min\{\rho^1, d^1\} & > & 1, \\ \rho^1 + 2^{-1}\min\{\rho_1, d_1\} & > & 1. \end{array}\right\} \qquad (3.9)$$

Then the wave operators $W^{\pm}(H, H_0)$ *exist and are complete. Moreover, the same conclusion holds if* V *is a finite sum of functions satisfying the condition above (for different* X_1, X^1, ρ_1 *and* ρ^1*) and of a function satisfying* (1.15) *with* $\rho > 1$.

A proof of this result follows the scheme of the proof of Theorem 2.4. However the role of the Sobolev trace theorem is played by the following analytical assertion [102].

Theorem 3.4 *Let* $\mathsf{H}^{l_1,l^1}(\mathbf{R}^d)$ *consist of functions* g *such that*

$$||g||^2_{l_1,l^1} = \int_{\mathbf{R}^d} |\check{g}(x_1,x^1)|^2(1+|x_1|^2)^{l_1}(1+|x^1|^2)^{l^1}dx < \infty, \quad \check{g} = \mathcal{F}^* g.$$

Then functions $g_\lambda(\omega) = g(\lambda^{1/2}\omega)$ *satisfy the estimate*

$$||g_\lambda||_{\mathfrak{N}} \le C||g||_{l_1,l^1}$$

if (and only if)

$$2l_1 + \min\{l^1, d^1/2\} > 1, \quad 2l^1 + \min\{l_1, d_1/2\} > 1.$$

Moreover, the functions g_λ *depend Hölder continuously in* $\mathfrak{N}$ *on* $\lambda > 0$.

By Theorem 3.2 for potentials obeying estimate (3.8), the wave operators $W^\pm(H, H_0)$ exist if $\rho_1+\rho^1 > 1$. However as we shall see in subsection 15.2, their completeness breaks down if one of inequalities (3.9) is not fulfilled.

3.4. The sharp form of the limiting absorption principle. The first estimate (2.1) and Theorem 2.8 (the limiting absorption principle) for the operator H_0 are optimal in the scale of Sobolev spaces (these results are violated for $l = 1/2$) but can be improved in terms of Besov spaces. More precisely, let the space $\mathbf{B}$ consist of functions f such that the norm

$$||f||_B = \left(\int_{|x|\le 1} |f(x)|^2dx\right)^{1/2} + \sum_{n=0}^{\infty}\left(2^n \int_{2^n\le|x|\le 2^{n+1}} |f(x)|^2dx\right)^{1/2} < \infty.$$

The space $\mathbf{B}^*$, dual to $\mathbf{B}$ with respect to $L_2 = L_2(\mathbf{R}^d)$, is the Banach space with one of its equivalent norms given by

$$||f||_{\mathbf{B}^*} = \sup_{r\ge 1}\left(r^{-1}\int_{|x|\le r} |f(x)|^2dx\right)^{1/2}. \tag{3.10}$$

The closure $\mathbf{B}^*_0$ of L_2 in the norm of $\mathbf{B}^*$ consists of functions $f(x)$ satisfying

$$\lim_{r\to\infty} r^{-1}\int_{|x|\le r} |f(x)|^2dx = 0.$$

Clearly,

$$L_2^{(l)} \subset \mathbf{B} \subset L_2^{(1/2)} \subset L_2 \subset L_2^{(-1/2)} \subset \mathbf{B}^*_0 \subset \mathbf{B}^* \subset L_2^{(-l)}, \quad \forall l > 1/2.$$

We remark that $f \in \mathbf{B}$ or $f \in \mathbf{B}^*$ if and only if its Fourier transform $\hat{f}$ belongs to the Besov space $\mathsf{B}^{1/2}_{2,1}$ or $\hat{f} \in \mathsf{B}^{-1/2}_{2,\infty}$, respectively. Of course, the functions

$$|x|^{-(d-1)/2}a(\hat{x})\exp(\pm i\lambda^{1/2}|x|), \quad \forall a \in L_2(\mathbf{S}^{d-1}),$$

belong to the space $\mathbf{B}^*$ but not to $\mathbf{B}^*_0$.

The following result of [4] complements Theorem 2.8.

Theorem 3.5 *Under the assumptions of Theorem* 2.5, *the norm of the operator* $R(z) : \mathbf{B} \to \mathbf{B}^*$ *is uniformly bounded:*

$$||R(z)||_{\mathbf{B},\mathbf{B}^*} \leq C, \tag{3.11}$$

where the constant C *does not depend on* z *as long as* $0 < \lambda_0 \leq \operatorname{Re} z \leq \lambda_1 < \infty$, $\operatorname{Im} z \neq 0$. *Moreover, for any* $f, g \in \mathbf{B}$, *the function* $(R(z)f, g)$ *is continuous with respect to* z *up to the cut along* $[0, \infty)$ *with exception of the point* $z = 0$.

Theorem 3.5 can first be checked for the free resolvent $R_0(z)$ and then, similarly to subsection 2.1, extended to the general case with the help of the resolvent identity (1.13).

Inequality (3.11) for a fixed $z = \lambda \pm i0$ leads to the high energy estimate of the resolvent. We discuss only the free case $V = 0$.

Corollary 3.6 *If* $\lambda \to \infty$, *then*

$$||R_0(\lambda \pm i0)||_{\mathbf{B},\mathbf{B}^*} = O(\lambda^{-1/2}). \tag{3.12}$$

Indeed, let $\mathbf{G}_\alpha$ be a dilation operator defined by

$$(\mathbf{G}_\alpha f)(x) = \alpha^{-d/2} f(\alpha^{-1} x). \tag{3.13}$$

Making the change of variables $x \mapsto \alpha x$ in (3.10), we see that

$$||\mathbf{G}_{\alpha^{-1}} g||_{\mathbf{B}^*} \leq C\alpha^{1/2} ||g||_{\mathbf{B}^*}, \quad \alpha \geq \alpha_0 > 0,$$

and, by duality,

$$||\mathbf{G}_\alpha f||_{\mathbf{B}} \leq C\alpha^{1/2} ||f||_{\mathbf{B}}, \quad \alpha \geq \alpha_0 > 0. \tag{3.14}$$

Since $H_0 \mathbf{G}_\alpha = \alpha^{-2} \mathbf{G}_\alpha H_0$, it follows from Theorem 3.5 that for any $f, g \in \mathbf{B}$

$$|(R_0(\lambda \pm i0)f, g)| = \lambda^{-1} |(R_0(1 \pm i0)\mathbf{G}_{\lambda^{1/2}} f, \mathbf{G}_{\lambda^{1/2}} g)| \leq C\lambda^{-1} ||\mathbf{G}_{\lambda^{1/2}} f||_{\mathbf{B}} \, ||\mathbf{G}_{\lambda^{1/2}} g||_{\mathbf{B}}.$$

By virtue of (3.14) the right-hand side here is bounded by $\lambda^{-1/2} ||f||_{\mathbf{B}} \, ||g||_{\mathbf{B}}$, which proves (3.12).

Corollary 3.7 *The operators* $\Gamma_0(\lambda) : \mathbf{B} \to \mathfrak{N}$ *and* $\Gamma_0(\lambda)^* : \mathfrak{N} \to \mathbf{B}^*$ *defined by* (1.21) *and* (2.11), *respectively, are bounded uniformly in* λ *for* $\lambda \geq \lambda_0 > 0$ (*and their norms are* $O(\lambda^{-1/4})$ *as* $\lambda \to \infty$).

Estimate (3.11) is optimal since, according to (1.20), even for $f \in C_0^\infty(\mathbf{R}^d)$, the function $R_0(\lambda \pm i0)f \in \mathbf{B}^*$ but does not belong even to the space $\mathbf{B}_0^*$. Theorem 3.5 allows us to extend asymptotics (1.20) to arbitrary $f \in \mathbf{B}$.

Proposition 3.8 *For any* $f \in \mathbf{B}$,

$$(R_0(\lambda \pm i0)f)(x) = c_\pm(\lambda)(\Gamma_0(\lambda)f)(\pm\hat{x})|x|^{-(d-1)/2} e^{\pm i\lambda^{1/2}|x|} + \varepsilon(x),$$

where $\varepsilon \in \mathbf{B}_0^*$.

3.5. Eigenfunctions of the continuous spectrum. Definition (1.18) of solutions $\psi^{\pm}(x;\omega,\lambda)$ of the Schrödinger equation makes sense if a potential $V(x)$ obeys (1.15) with some $\rho > (d+1)/2$. Indeed, according to the limiting absorption principle, the term

$$R(\lambda \mp i0)V\psi_0(\omega,\lambda) = (R(\lambda \mp i0)\langle x\rangle^{-l})(\langle x\rangle^{l}V\psi_0(\omega,\lambda)), \quad l \in (1/2, \rho - d/2),$$

is well-defined as an element of the space $\mathbf{B}^*$, and it satisfies the Lippmann-Schwinger equation (1.19). Finally, it follows from Proposition 3.8 that the function $\psi = \psi^-$ has asymptotics (1.16) as $|x| \to \infty$ in the following sense.

Theorem 3.9 *Suppose that estimate* (1.15) *holds for some* $\rho > (d+1)/2$. *Then for any* $\lambda > 0$ *and any unit vector* $\omega \in \mathbb{S}^{d-1}$ *equation* (1.14) *has a solution with the asymptotics*

$$\psi(x;\omega,\lambda) = \exp(i\lambda^{1/2}\langle\omega,x\rangle) + a(\hat{x},\omega;\lambda)|x|^{-(d-1)/2}\exp(i\lambda^{1/2}|x|) + \varepsilon(x), \quad \varepsilon \in \mathbf{B}_0^*, \tag{3.15}$$

as $|x| \to \infty$. *The scattering amplitude* $a(\hat{x},\omega;\lambda)$ *belongs to the space* $L_2(\mathbb{S}^{d-1})$ *in the variable* $\hat{x}$ *uniformly in* $\omega \in \mathbb{S}^{d-1}$.

We emphasize that a term $\varepsilon \in \mathbf{B}_0^*$ satisfies the estimate $o(|x|^{-(d-1)/2})$ in an averaged sense.

In the general case $\rho > 1$ eigenfunctions of the continuous spectrum of the Schrödinger operator H were constructed in subsection 2.3 by formula (2.18). According to Theorem 3.5 and Corollary 3.7, the operators $\Gamma^{\pm}(\lambda) : \mathbf{B} \to \mathfrak{N}$ and $\Gamma^{\pm}(\lambda)^* : \mathfrak{N} \to \mathbf{B}^*$ are bounded. Consequently, functions (2.18) belong to the space $\mathbf{B}^*$. Here we shall find their asymptotics as $|x| \to \infty$. Actually, we obtain asymptotics of an arbitrary solution of the Schrödinger equation from the class $\mathbf{B}^*$. This implies, in particular, that all such solutions are parametrized by functions $a \in \mathfrak{N} = L_2(\mathbb{S}^{d-1})$.

The following result was obtained in the paper [163] where the technique of [4] and [65] (in particular, Theorem 3.5) was essentially used.

Theorem 3.10 *Suppose that estimate* (1.15) *holds for some* $\rho > 1$. *Let* $u \in \mathbf{B}^*$ *be a solution of equation* (1.14). *Then there exist functions* $a_{\pm} \in \mathfrak{N}$ *such that*

$$u(x) = |x|^{-(d-1)/2}\Big(\gamma a_+(\hat{x})\exp(i\lambda^{1/2}|x|) - \bar{\gamma}a_-(-\hat{x})\exp(-i\lambda^{1/2}|x|)\Big) + \varepsilon(x), \tag{3.16}$$

where $\gamma = \exp(-i\pi(d-3)/4)$ *and* $\varepsilon \in \mathbf{B}_0^*$. *Functions* $a_{\pm}$ *in* (3.16) *are related by the scattering matrix* : $a_+ = S(\lambda)a_-$.

Conversely, for every $a_+ \in \mathfrak{N}$ (*or* $a_- \in \mathfrak{N}$) *there exists a unique function* $a_- \in \mathfrak{N}$ (*or* $a_+ \in \mathfrak{N}$) *and a unique solution of equation* (1.14) *satisfying condition* (3.16) *with* $\varepsilon \in \mathbf{B}_0^*$.

The numerical coefficient γ is introduced in (3.16) in order that $a_+ = a_-$ for $V = 0$. Asymptotics (3.15) and (3.16) can be differentiated in the variable $r = |x|$. By virtue of the following assertion (see e.g. [136]) this determines uniquely solutions ψ and u of equation (1.14).

Theorem 3.11 *Let estimate* (1.15) *hold for some* $\rho > 1$. *Suppose that a function* $u \in \mathsf{H}^2_{loc}(\mathbb{R}^d)$ *satisfies equation* (1.14) *and*

$$||u(r_n\cdot)||_{\mathfrak{N}} = O(r_n^{-(d-1)/2}), \quad ||((\partial_r \mp i\lambda^{1/2})u)(r_n\cdot)||_{\mathfrak{N}} = o(r_n^{-(d-1)/2})$$

for some sequence $r_n \to \infty$ *and one of the signs* " + " *or* " − ". *Then* $u = 0$.

The solutions of the Schrödinger equation with asymptotics (3.16) were of course extensively discussed in the physics literature [112, 125]. We remark also that in the recent paper [117] the existence of solutions of the corresponding differential equation with asymptotics similar to (3.16) was established for $a_\pm \in C^\infty(\mathbb{S}^{d-1})$ in the general context of asymptotically Euclidean manifolds.

Finally, we mention the paper [145] devoted to construction of solutions of the Schrödinger equation with asymptotics of type (1.16) at infinity. The method of this paper is quite different from the one exposed above and requires that

$$\partial^\kappa V(x) = O(|x|^{-\rho-|\kappa|}) \tag{3.17}$$

for all κ but ρ may be an arbitrary number bigger that 1.

3.6. Standing waves. Let $\mathcal{J}$ be the reflection operator on the unit sphere, i.e.,

$$(\mathcal{J}f)(\omega) = f(-\omega). \tag{3.18}$$

We shall see in Section 8 that, for any $\lambda > 0$, both operators $S = S(\lambda)$ and the modified scattering matrix

$$\Sigma = \Sigma(\lambda) = S(\lambda)\mathcal{J} \tag{3.19}$$

have pure point spectrum. If $\Sigma(\lambda)a = \exp(2i\theta)a$, then it follows from Theorem 3.10 that equation (1.14) has a solution with asymptotics of the standing wave as $|x| \to \infty$:

$$u(x) = a(\hat{x})|x|^{-(d-1)/2}\sin(\lambda^{1/2}|x| - \pi(d-3)/4 + \theta) + \varepsilon(x), \tag{3.20}$$

where $\varepsilon \in \mathbf{B}_0^*$.

In the spherically symmetric case $V(x) = V(|x|)$, solutions with asymptotics (3.20) are naturally parametrized by the orbital quantum number $m = 0, 1, 2, \ldots$. In this case $a = a_m$ is the spherical function and $\delta_m = \theta_m + \pi m/2$ is the limiting (or scattering) phase (see e.g. [112]) defined by the asymptotics

$$\psi_m(r,\lambda) = C_m(\lambda)\sin\bigl(\lambda^{1/2}r - \pi(2m+d-3)/4 + \delta_m\bigr) + o(1) \tag{3.21}$$

as $r \to \infty$ of the regular (such that $\psi_m(r,\lambda) \sim r^{m+(d-1)/2}$ as $r \to 0$) solution of the Schrödinger equation

$$-\psi_m'' + p_m r^{-2}\psi_m + V(r)\psi_m = \lambda\psi_m, \quad p_m = (m + (d-2)/2)^2 - 1/4. \tag{3.22}$$

4. LONG-RANGE INTERACTIONS. THE SCHEME OF SMOOTH PERTURBATIONS

Using the smooth method, we give here a stationary proof of completeness of (modified) wave operators for the Schrödinger operator $H = -\Delta + V$ with a long-range potential V.

4.1. Modified wave operators. The condition of Theorem 2.4 is optimal even for the existence of wave operators. Actually, if a potential $V(x)$ satisfies estimate (1.15) for $\rho \le 1$, then the wave operators (1.6) for the pair $H_0 = -\Delta$, $H = -\Delta + V$ do not in general exist (for example, for the Coulomb potential $V(x) = v|x|^{-1}$). Nevertheless the asymptotic behaviour of the function $\exp(-iHt)f$ for large $|t|$ remains sufficiently close to the free evolution $\exp(-iH_0t)f_0$ if the condition

$$|\partial^\kappa V(x)| \le C_\kappa(1+|x|)^{-\rho-|\kappa|}, \quad \rho > 0, \tag{4.1}$$

is satisfied for $|\kappa| \le \kappa_0$ with κ_0 big enough. Potentials obeying this condition for some $\rho \in (0,1]$ are called long-range.

There are several possible descriptions of $\exp(-iHt)f$ as $t \to \pm\infty$. One of them is a modification of the free evolution which, in its turn, can be done either in momentum [42, 29, 66] or in coordinate [158] representations. Here we discuss the coordinate modification. Set

$$(U_0(t)f)(x) = \exp(i\Xi(x,t))(2it)^{-d/2}\hat{f}(x/(2t)), \tag{4.2}$$

where $\hat{f} = \mathcal{F}_0 f$ is the Fourier transform of f. Recall (see subsection 3.2) that for short-range potentials V, when

$$\Xi(x,t) = (4t)^{-1}|x|^2,$$

the wave operators $W^\pm(H, H_0)$ and (3.5) coincide. In the long-range case the limit (3.5) still exists if the function $\Xi(x,t)$ is a (perhaps, approximate) solution of the eikonal equation

$$\partial\Xi/\partial t + |\nabla\Xi|^2 + V = 0. \tag{4.3}$$

Actually, we seek $\Xi(x,t)$ in the form

$$\Xi(x,t) = (4t)^{-1}|x|^2 + \Omega(x,t) \tag{4.4}$$

and require that $\Omega(x,t)$ satisfies for any $c_1, c_2 > 0$ the estimates

$$\sup_{c_1|t|\le|x|\le c_2|t|} |(\partial_x^\kappa\Omega)(x,t)| \le C|t|^{1-|\kappa|-\varepsilon}, \quad |\kappa| = 1,2, \quad \varepsilon > 0, \tag{4.5}$$

and

$$\sup_{c_1|t|\leq|x|\leq c_2|t|} \left|\Omega_t + t^{-1}\langle x, \nabla\Omega\rangle + |\nabla\Omega|^2 + V(x)\right| \leq C|t|^{-1-\varepsilon}, \quad \varepsilon > 0. \tag{4.6}$$

For example, if $\rho > 1/2$ we can neglect the nonlinear term $|\nabla\Omega|^2$ in (4.6) and set

$$\Xi(x,t) = (4t)^{-1}|x|^2 - t\int_0^1 V(sx)ds. \tag{4.7}$$

In the case $\rho \leq 1/2$, one has to apply the method of successive approximations to (4.3) and to keep $[\rho^{-1}]$ terms. In particular, if $\rho \in (1/4, 1/2]$, only one term

$$-t^3 \int_0^1 s^2 \left|\int_0^1 \sigma(\nabla V)(\sigma x)d\sigma\right|^2 ds$$

should be added to the right-hand side of (4.7). With the phase $\Xi(x,t)$ constructed in such a way, for an arbitrary $\hat{f} \in C_0^\infty(\mathbf{R}^d \setminus \{0\})$, the function $U_0(t)f$ is an approximate solution of the Schrödinger equation, that is condition (3.6) holds. As in subsection 3.2, this implies the following

Theorem 4.1 *Under assumption* (4.1), *the wave operators* (3.5) *exist.*

Note also that, similarly to subsection 3.2, it suffices to require (4.1) in any closed cone not intersecting some closed set $N \subset \mathbf{S}^{d-1}$ of measure zero. Under the assumption of their existence, the wave operators (3.5) are automatically isometric. Furthermore, it follows from (4.5), (4.6) that for any $s \in \mathbf{R}$

$$s - \lim_{|t|\to\infty} U_0^*(t+s)U_0(t) = \exp(iH_0 s) \tag{4.8}$$

and hence operators (3.5) enjoy the intertwining property $H\mathcal{W}^\pm = \mathcal{W}^\pm H_0$. As in the short-range case, the wave operator is said to be complete if

$$\operatorname{Ran} \mathcal{W}^\pm = \mathcal{H}^{(ac)}.$$

Only the completeness of $\mathcal{W}^\pm$ is a non-trivial mathematical problem.

Of course, the function $\Xi(x,t)$ in definition (4.2) of a modified free evolution is not defined uniquely. An arbitrary (smooth) function $\theta_\pm(x/(2t))$ can be added to $\Xi(x,t)$ for $\pm t > 0$. Then the wave operator $\mathcal{W}^\pm$ is replaced by $\mathcal{W}^\pm \mathcal{F}^* e^{i\theta_\pm(\xi)}\mathcal{F}$.

4.2. The limiting absorption principle and radiation estimates. The starting point of the stationary approach to the completeness of modified wave operators is the limiting absorption principle.

Theorem 4.2 *Let a potential* $V = V_S + V_L$ *where a short-range term* V_S *satisfies* (1.15) *for some* $\rho = \rho_0 > 1$ *and a long-range term* V_L *satisfies the estimate*

$$|V_L(x)| + |x|\,|\partial V_L(x)/\partial|x|\,| \leq C(1+|x|)^{-\rho_1}, \quad \rho_1 > 0.$$

Then for any $l > 1/2$, *the operator-function* $\langle x\rangle^{-l}R(z)\langle x\rangle^{-l}$ *is norm-continuous in* z *in the region* $\operatorname{Re} z > 0$, $\pm\operatorname{Im} z \geq 0$. *In particular, the operator* $\langle x\rangle^{-l}$ *is* H*-smooth on any compact interval* $\Lambda \subset (0,\infty)$, *and the positive spectrum of the operator* H *is absolutely continuous. Moreover, the conclusion of Theorem* 3.5 *also holds.*

In the long-range case perturbative arguments of Section 2 do not work. The simplest proof of the limiting absorption principle hinges on the Mourre estimate (see [120] and the books [34], [5]) for the commutator

$$i([H,\mathbf{A}]u,u) \geq c\|u\|^2, \quad c=c_\lambda>0, \quad u\in E(\Lambda_\lambda)\mathcal{H}, \tag{4.9}$$

where

$$\mathbf{A}=\sum(x_jD_j+D_jx_j), \quad D_j=-i\partial_j, \quad j=1,\dots,d, \tag{4.10}$$

is the generator of dilations, $\lambda>0$ is not an eigenvalue of the operator H and $\Lambda_\lambda \ni \lambda$ is a sufficiently small interval.

Under the assumptions of Theorem 4.2 the proof of (4.9) is quite elementary. One first remaks that

$$i[H_0,\mathbf{A}]=4H_0 \tag{4.11}$$

(this is an analog of the canonical commutation relation $i[D_j,x_j]=I$). To estimate the commutator $[V,\mathbf{A}]$, we consider the long-range V_L and short-range V_S parts of V separately. A direct calculation shows that

$$i[V_L,\mathbf{A}]=-2\langle x,\nabla V_L\rangle=-2|x|\partial V_L(x)/\partial|x|.$$

We do not really commute V_S with $\mathbf{A}$ using only that

$$|(\mathbf{A}u,V_Su)|\leq C(\|\langle x\rangle^{-\varepsilon_0}\nabla u\|^2+\|\langle x\rangle^{-\varepsilon_0}u\|^2), \quad 2\varepsilon_0=\rho_0-1.$$

Replacing in the right-hand side of (4.11) H_0 by $H-V$, we see that

$$i([H,\mathbf{A}]u,u)\geq 4(Hu,u)-c\|\langle x\rangle^{-\varepsilon}u\|^2-c\|\langle x\rangle^{-\varepsilon_0}\nabla u\|^2, \quad \varepsilon=\min\{\varepsilon_0,\rho_1/2\}. \tag{4.12}$$

Let us apply (4.12) to elements $u=E(\Lambda_\lambda^{(\delta)})u$ where $\Lambda_\lambda^{(\delta)}=(\lambda-\delta,\lambda+\delta)$. If $\lambda\notin\sigma_H^{(p)}$, then $E(\Lambda_\lambda^{(\delta)})\to 0$ strongly as $\delta\to 0$. Since the operators $\langle x\rangle^{-\varepsilon}E(X)$ and $\langle x\rangle^{-\varepsilon_0}\nabla E(X)$ are compact for any bounded set X, we have that

$$\lim_{\delta\to 0}\|\langle x\rangle^{-\varepsilon}E(\Lambda_\lambda^{(\delta)})\|=0, \quad \lim_{\delta\to 0}\|\langle x\rangle^{-\varepsilon_0}\nabla E(\Lambda_\lambda^{(\delta)})\|=0.$$

Taking into account that

$$HE(\Lambda_\lambda^{(\delta)})\geq(\lambda-\delta)E(\Lambda_\lambda^{(\delta)})\geq cE(\Lambda_\lambda^{(\delta)}), \quad c>0,$$

if $\lambda>0$ and δ is small enough, we arrive at (4.9).

The Mourre estimate implies that positive eigenvalues of H may accumulate at zero and infinity only and the operator-function $\langle x\rangle^{-l}R(z)\langle x\rangle^{-l}$ is norm-continuous in z in the region $\operatorname{Re} z>0$, $\pm\operatorname{Im} z\geq 0$ except eigenvalues of H. These results are deduced from estimate (4.9) by arguments which are actually of abstract nature (see [120, 34, 5]). As shown in [91], in the long-range case the sharp form (3.11) of the limiting absorption principle is also true.

Note however that the Mourre approach does not exclude the existence of the discrete set of positive eigenvalues of the operator H. Another approach to a proof of Theorem 4.2 (see, e.g., [136]) has a partial differential equations flavour rather than that of functional analysis. Its advantage is that it proves also the absence of positive eigenvalues of the operator H.

In the short-range case the limiting absorption principle suffices (see subsection 2.1) for construction of scattering theory but, for long-range potentials, one needs an additional analytical information pertaining in some sense to the critical case $l = 1/2$. However, as seen from asymptotics (1.20), the operator $\langle x\rangle^{-1/2}$ is not even H_0-smooth. Let $(\nabla^{\perp}u)(x)$ be the orthogonal projection of the gradient $(\nabla u)(x)$ on the plane orthogonal to the vector x. The following result (appeared first in [166] in the N-particle framework) shows that the differential operator $\nabla^{\perp}$ improves the fall-off of functions $(U(t)f)(x)$ for large t and x.

Theorem 4.3 *Let*

$$(\nabla_j^{\perp}u)(x) = (\partial_j u)(x) - |x|^{-2}\langle(\nabla u)(x), x\rangle x_j, \quad j = 1, \ldots, d. \tag{4.13}$$

Under the assumptions of Theorem 4.2, the operators $G_j = \langle x\rangle^{-1/2}\nabla_j^{\perp}$ are H-smooth on any compact interval $\Lambda \subset (0, \infty)$.

We refer to the estimates of Theorem 4.3 as radiation estimates. Similar (but not the same – see Theorem 5.2 below) estimates can be found, for example, in [136] where their proof is intimately connected to those of the limiting absorption principle and of the absence of positive eigenvalues.

Our proof of Theorem 4.3 relies on Proposition 1.19. Let

$$M = \sum(m^{(j)}D_j + D_j m^{(j)}), \quad m^{(j)} = \partial m/\partial x_j, \tag{4.14}$$

where $m \in C^{\infty}(\mathbf{R}^d)$ and $m(x) = |x|$ for $|x| \geq 1$. To calculate the commutator $[H, M]$, we remark first that for any function m

$$i[H_0, M] = 4\sum_{j,k} D_j m^{(jk)} D_k - (\Delta^2 m), \quad m^{(jk)} = \partial^2 m/\partial x_j \partial x_k. \tag{4.15}$$

In particular, if $m(x) = |x|$, then

$$\sum_{j,k} m^{(jk)} u_j \bar{u}_k = |x|^{-1}|\nabla^{\perp}u|^2 = |x|^{-1}\sum_{j=1}^{d}|\nabla_j^{\perp}u|^2, \quad u_j = \partial u/\partial x_j, \tag{4.16}$$

and $(\Delta^2 m)(x) = O(|x|^{-3})$ as $|x| \to \infty$.

Our estimate of the commutator $[V, M]$ is similar to that of $[V, \mathbf{A}]$. For the long-range part V_L, we have that

$$[V_L, M] = 2i\langle\nabla m, \nabla V_L\rangle. \tag{4.17}$$

If $m(x) = |x|$, then

$$\langle\nabla m, \nabla V_L\rangle = \partial V_L(x)/\partial|x| = O(|x|^{-1-\rho_1}).$$

We again do not really commute the short-range part V_S with M using only that

$$|(Mu, V_S u)| \leq C(||\langle x\rangle^{-\rho_0/2}\nabla u||^2 + ||\langle x\rangle^{-\rho_0/2}u||^2)$$
$$\leq C_1||\langle x\rangle^{-\rho_0/2}(H_0+I)^{1/2}u||^2. \tag{4.18}$$

It follows that

$$|([V, M]u, u)| \leq C||\langle x\rangle^{-l}(H_0+I)^{1/2}u||^2, \tag{4.19}$$

where

$$2l = \min\{\rho_0, \rho_1 + 1\} > 1. \tag{4.20}$$

Combining (4.15), (4.16) and (4.19), we arrive at the estimate

$$i([H, M]u, u) \geq 4||\langle x\rangle^{-1/2}\nabla^\perp u||^2 - c^2||\langle x\rangle^{-l}(H_0+I)^{1/2}u||^2.$$

This is exactly estimate (1.32) for $G = 2\langle x\rangle^{-1/2}\nabla^\perp$, $K = c\langle x\rangle^{-l}(H_0+I)^{1/2}$ and H-bounded operator (4.14). Now let us take into account that, by Theorem 4.2, the operator $\langle x\rangle^{-l}(H_0+I)^{1/2}E(\Lambda)$ is H-smooth (the factor $(H_0+I)^{1/2}$ is compensated here by $E(\Lambda)$). Thus Proposition 1.19 implies that the operator $\langle x\rangle^{-1/2}\nabla^\perp E(\Lambda)$ is also H-smooth.

4.3. Stationary scattering theory. Long-range scattering theory can be conveniently formulated in terms of wave operators (1.8). It is natural [82, 83] to choose $J = J^\pm$ as a pseudo-differential operator (PDO)

$$(J^\pm f)(x) = (2\pi)^{-d/2}\int_{\mathbf{R}^d} e^{i\langle x,\xi\rangle}\mathrm{j}^\pm(x,\xi)\hat{f}(\xi)d\xi \tag{4.21}$$

with a suitable symbol $\mathrm{j}^\pm(x,\xi)$ depending on a long-range potential V and on the sign of t.

Here we outline a proof (see [172], for details) of the existence and completeness of wave operators $W^\pm(H, H_0, J^\pm)$ for the pair $H_0 = -\Delta, H = -\Delta + V$. Since this proof relies on the theory of PDO, we suppose that V satisfies condition (4.1) for all κ. Moreover, to simplify our presentation, we assume that $\rho \in (1/2, 1]$. We consider the class $S^m_{\rho,\delta}$ of symbols (introduced by L. Hörmander) $a \in C^\infty(\mathbf{R}^d \times \mathbf{R}^d)$ obeying the estimates

$$|\partial_x^\alpha \partial_\xi^\beta a(x,\xi)| \leq C_{\alpha,\beta}(1+|x|)^{m-\rho|\alpha|+\delta|\beta|}$$

for some numbers $\rho, \delta \in [0,1]$ and all multi-indices α, β. We assume also that $a(x,\xi) = 0$ for sufficiently large $|\xi|$. Recall that the usual PDO calculus (see e.g. v.3 of [67] or [138]) requires the condition $\rho > 1/2 > \delta$. Compared to this calculus, in scattering theory the roles of the variables x and ξ are interchanged.

We need the following elementary facts. Let A be a PDO with symbol $a \in S^m_{\rho,\delta}$. Then the operator $A\langle x\rangle^{-m}$ is bounded and $A\langle x\rangle^{-m_1}$, $m_1 > m$, is compact in the space $L_2(\mathbf{R}^d)$. The adjoint A^* to A is also a PDO, whose symbol equals $\overline{a(x,\xi)}$, up to a term from the class $S^{m-\rho+\delta}_{\rho,\delta}$. If A_j, $j = 1, 2$, are PDO with symbols $a_j \in S^{m_j}_{\rho,\delta}$, then the product $A_1 A_2$ is a PDO, whose symbol equals $a_1(x,\xi)a_2(x,\xi)$, up to a term from the class $S^{m_1+m_2-\rho+\delta}_{\rho,\delta}$.

Our goal is to find a PDO (4.21) such that the perturbation

$$T^{\pm} = HJ^{\pm} - J^{\pm}H_0$$

admits a factorization into a product of H_0- and H-smooth operators, and thus both triples $H_0, H, J^{\pm}$ fit into the framework of the smooth perturbations theory. Roughly speaking, this means that $e^{i\langle x,\xi\rangle}\mathbf{j}^{\pm}(x,\xi)$ are approximate eigenfunctions of the operator H corresponding to "eigenvalues" $|\xi|^2$. Of course, $T^{\pm}$ is also a PDO with symbol

$$t^{\pm}(x,\xi) = (-\Delta + V(x) - |\xi|^2)\mathbf{j}^{\pm}(x,\xi). \tag{4.22}$$

Let us first try to seek $\mathbf{j}^{\pm}(x,\xi)$ in the form $\mathbf{j}^{\pm}(x,\xi) = \exp(i\Phi^{\pm}(x,\xi))$. Set

$$\varphi^{\pm}(x,\xi) = \langle \xi, x\rangle + \Phi^{\pm}(x,\xi). \tag{4.23}$$

The expression

$$(-\Delta + V - |\xi|^2)(e^{i\varphi^{\pm}}) = e^{i\varphi^{\pm}}(2\langle \xi, \nabla\Phi^{\pm}(x,\xi)\rangle + |\nabla\Phi^{\pm}|^2 - i\Delta\Phi^{\pm} + V(x)) \tag{4.24}$$

is "small" if $\Phi^{\pm}$ satisfies (perhaps, approximately) the equation

$$2\langle \xi, \nabla\Phi^{\pm}(x,\xi)\rangle + |\nabla\Phi^{\pm}(x,\xi)|^2 + V(x) = 0, \quad \nabla = \nabla_x, \tag{4.25}$$

(which is equivalent to the eikonal equation $|\nabla\varphi^{\pm}|^2 + V = |\xi|^2$ for $\varphi^{\pm}$) and $\Delta\Phi^{\pm}$ is a short-range function.

The notorious difficulty (for $d \geq 2$) of this approach is that equation (4.25) does not have global solutions (such that $|\nabla\Phi^{\pm}(x,\xi)| \to 0$ as $|x| \to \infty$). However it is easy to construct functions $\Phi^{\pm}$ satisfying approximately (4.25) if a conical neighbourhood of the direction $\hat{x} = \mp\hat{\xi}$ is removed from $\mathbf{R}^d$. In particular, in the case $\rho > 1/2$ the term $|\nabla\Phi^{\pm}(x,\xi)|^2$ can be neglected in (4.25), and an exact solution of the corresponding linearized equation

$$2\langle \xi, \nabla\Phi^{\pm}(x,\xi)\rangle + V(x) = 0$$

can be defined by the explicit formula

$$\Phi^{\pm}(x,\xi) = \pm 2^{-1}\int_0^{\infty}\Big(V(x \pm \tau\xi) - V(\pm\tau\xi)\Big)d\tau. \tag{4.26}$$

Clearly, for all multi-indices α, β and any $\nu \in (-1,1)$

$$|\partial_x^{\alpha}\partial_{\xi}^{\beta}\Phi^{\pm}(x,\xi)| \leq C_{\alpha,\beta,\nu}(1+|x|)^{1-\rho-|\alpha|}, \quad \pm\langle\hat{\xi},\hat{x}\rangle \geq \pm\nu, \quad |\alpha| \geq 1, \tag{4.27}$$

so (4.25) is satisfied up to a term $O(|x|^{-2\rho})$.

Now we can define a symbol of the PDO $J^{\pm}$. Let $\sigma^{\pm} \in C^{\infty}(-1,1)$ be such that $\sigma^{\pm}(\vartheta) = 1$ in a neighbourhood of the point ± 1 and $\sigma^{\pm}(\vartheta) = 0$ in a neighbourhood of the point ∓ 1, let $\eta \in C^{\infty}(\mathbf{R}^d)$ be such that $\eta(x) = 0$ in a neighbourhood of zero and $\eta(x) = 1$ for large $|x|$ and let $\psi \in C_0^{\infty}(\mathbf{R}_+)$. We set

$$\mathbf{j}^{\pm}(x,\xi) = e^{i\Phi^{\pm}(x,\xi)}\eta(x)\sigma^{\pm}(\langle\hat{\xi},\hat{x}\rangle)\psi(|\xi|^2). \tag{4.28}$$

Due to the function $\psi(|\xi|^2)$ all our considerations will be localized on a bounded energy interval disjoint from zero. The function η is introduced only to get rid of the singularity of the function $|x|^{-1}x$ at the point $x = 0$. The most important cut-off $\sigma^\pm$ restricts the symbol $\mathrm{j}^\pm(x,\xi)$ on the region where estimate (4.27) is satisfied. However a price to pay for this cut-off is that due to $\sigma^\pm$ symbol (4.22) decays as $|x|^{-1}$ only at infinity. According to (4.27), $\mathrm{j}^\pm \in \mathcal{S}^0_{\rho,\delta}$, where ρ is the same number as in (4.1) and $\delta = 1 - \rho$. In particular, $J^\pm$ is a bounded operator in the space $L_2(\mathbb{R}^d)$.

By virtue of (4.24) and (4.28), symbol (4.22) of the PDO $T^\pm$ equals

$$t^\pm(x,\xi) = e^{i\Phi^\pm(x,\xi)}(\tau^\pm_s(x,\xi) + \tau^\pm_r(x,\xi))\psi(|\xi|^2) =: t^\pm_s(x,\xi) + t^\pm_r(x,\xi). \tag{4.29}$$

The singular term

$$\tau^\pm_s(x,\xi) = -2i\eta(x)\langle\xi, \nabla\sigma^\pm(\langle\hat\xi,\hat x\rangle)\rangle = -2i\eta(x)\,|\xi|\,|x|^{-1}(1 - \langle\hat\xi,\hat x\rangle^2)(\sigma^\pm)'(\langle\hat\xi,\hat x\rangle) \tag{4.30}$$

decays as $|x|^{-1}$ as $|x| \to \infty$. The regular one

$$\tau^\pm_r = (|\nabla\Phi^\pm|^2 - i\Delta\Phi^\pm)\zeta^\pm - 2i\sigma^\pm\langle\xi,\nabla\eta\rangle - 2i\langle\nabla\Phi^\pm,\nabla\zeta^\pm\rangle - \Delta\zeta^\pm,$$

where $\zeta^\pm(x,\xi) = \eta(x)\sigma^\pm(\langle\hat\xi,\hat x\rangle)$. It follows from (4.27) that

$$|\partial^\alpha_x\partial^\beta_\xi\tau^\pm_r(x,\xi)| \le C_{\alpha,\beta}(1 + |x|)^{-2\rho-|\alpha|}. \tag{4.31}$$

We shall show that

$$T^\pm = \sum_{j=1}^{d} G^*_j B^\pm_s G_j + \langle x\rangle^{-l}B^\pm_r\langle x\rangle^{-l}, \tag{4.32}$$

where $G_j = \langle x\rangle^{-1/2}\nabla^\perp_j$, $l = \rho > 1/2$ and the operators $B^\pm_s, B^\pm_r$ are bounded. Let $T^\pm_s$ and $T^\pm_r$ be PDO with symbols $t^\pm_s$ and $t^\pm_r$ defined by (4.29), so that $T^\pm = T^\pm_s + T^\pm_r$. According to (4.31), the operator $\langle x\rangle^\rho T^\pm_r\langle x\rangle^\rho$ is bounded. To obtain (4.32) for the singular part $T^\pm_s$ of the perturbation $T^\pm$, let us define $B^\pm_s$ as a PDO with symbol

$$b^\pm_s(x,\xi) = (1 + |x|^2)^{1/2}\,|\xi|^{-2}(1 - \langle\hat\xi,\hat x\rangle^2)^{-1}t^\pm_s(x,\xi).$$

It follows from (4.29), (4.30) that

$$b^\pm_s(x,\xi) = -2i\eta(x)(1 + |x|^{-2})^{1/2}e^{i\Phi^\pm(x,\xi)}(\sigma^\pm)'(\langle\hat\xi,\hat x\rangle)|\xi|^{-1}\psi(|\xi|^2).$$

Hence $b^\pm_s \in \mathcal{S}^0_{\rho,\delta}$ and the operator $B^\pm_s$ is bounded. The principal symbol of the operator $\sum_{j=1}^d G^*_j B^\pm_s G_j$ equals

$$(1 + |x|^2)^{-1/2}\sum_{j=1}^{d}(\xi_j - |x|^{-2}\langle\xi,x\rangle x_j)^2 b^\pm_s(x,\xi) = t^\pm_s(x,\xi).$$

Therefore the symbol of the PDO $T^\pm_s - \sum_{j=1}^d G^*_j B^\pm_s G_j$ belongs to the class $\mathcal{S}^{-1-\rho}_{\rho,\delta}$. This implies representation (4.32).

By Theorems 4.2 and Theorem 4.3, the operators $\langle x\rangle^{-l}$, $l > 1/2$, and G_j are H_0- and H-smooth on any bounded disjoint from zero positive interval. So representation (4.32) plays the role of (1.31), and Proposition 1.18 yields the following result.

Lemma 4.4 *The wave operators*

$$W^{\pm}(H, H_0; J^{\pm}), \quad W^{\pm}(H_0, H; (J^{\pm})^*) \tag{4.33}$$

and

$$W^{\pm}(H, H_0; J^{\mp}), \quad W^{\pm}(H_0, H; (J^{\mp})^*) \tag{4.34}$$

exist. Operators (4.33) *as well as* (4.34) *are adjoint to each other.*

Below we fix a compact interval $\Lambda \subset (0, \infty)$ and choose a function $\psi \in C_0^\infty(\mathbb{R}_+)$ such that $\psi(\lambda) = 1$ on Λ.

Lemma 4.5 *The operators $W^{\pm}(H, H_0; J^{\pm})$ are isometric on the subspace $E_0(\Lambda)\mathcal{H}$ and $W^{\pm}(H, H_0; J^{\mp}) = 0$.*

Indeed, it suffices to check that

$$s - \lim_{t \to \pm\infty} ((J^{\pm})^* J^{\pm} - \psi^2(H_0))e^{-iH_0 t} = 0 \tag{4.35}$$

and

$$s - \lim_{t \to \pm\infty} (J^{\mp})^* J^{\mp} e^{-iH_0 t} = 0. \tag{4.36}$$

Up to a compact term, $(J^{\mp})^* J^{\mp}$ equals the PDO $Q^{\mp}$ with symbol $\zeta^{\mp}(x, \xi)^2 \psi^2(|\xi|^2)$. If $t \to \pm\infty$, then the stationary point $\xi = x/(2t)$ of the integral

$$(Q^{\mp} e^{-iH_0 t} f)(x) = (2\pi)^{-d/2} \eta^2(x) \int_{\mathbb{R}^d} e^{i\langle \xi, x\rangle - i|\xi|^2 t} \sigma^{\mp}(\langle \hat{\xi}, \hat{x}\rangle)^2 \psi^2(|\xi|^2) \hat{f}(\xi) d\xi. \tag{4.37}$$

does not belong to the support of the function $\sigma^{\mp}(\langle \hat{\xi}, \hat{x}\rangle)$. Therefore supposing that $f \in \mathcal{S}(\mathbb{R}^d)$ and integrating by parts, we estimate integral (4.37) by $C_N(1 + |x| + |t|)^{-N}$ for an arbitrary N. This proves (4.36). To prove (4.35), we apply the same arguments to the PDO with symbol $(\zeta^{\pm}(x, \xi)^2 - 1)\psi^2(|\xi|^2)$.

Lemma 4.6 *The operators $W^{\pm}(H_0, H; (J^{\pm})^*)$ are isometric on the subspace $E(\Lambda)\mathcal{H}$.*

In fact, by Lemma 4.5, $W^{\pm}(H, H_0; J^{\mp}) = 0$ and hence it follows from Lemma 4.4 that $W^{\pm}(H_0, H; (J^{\mp})^*) = 0$. This implies that

$$\lim_{t \to \pm\infty} ||(J^{\mp})^* e^{-iHt} f|| = 0, \quad f \in E(\Lambda)\mathcal{H}. \tag{4.38}$$

Let us choose the functions $\sigma^{\pm}$ in such a way that $\sigma^+(\vartheta)^2 + \sigma^-(\vartheta)^2 = 1$. Then, according to (4.28),

$$J^+(J^+)^* + J^-(J^-)^* = \psi^2(H) + K$$

for a compact operator K (we have taken into account here that $\psi^2(H) - \psi^2(H_0) \in \mathfrak{S}_\infty$), and (4.38) implies that

$$\lim_{t \to \pm\infty} ||(J^{\pm})^* e^{-iHt} f|| = ||f||.$$

This is equivalent to isometricity of the wave operator $W^{\pm}(H_0, H; (J^{\pm})^*)$.

Since both operators (4.33) are isometric, the asymptotic completeness

$$\operatorname{Ran} W^{\pm}(H, H_0; J^{\pm}) E_0(\Lambda) = E(\Lambda)\mathcal{H} \tag{4.39}$$

follows.

Theorem 4.7 *Suppose that condition* (4.1) *is fulfilled for* $\rho > 1/2$ *and let the operators* $J^\pm$ *be defined by* (4.21), (4.26) *and* (4.28). *Then the wave operators* (4.33) *exist, are isometric on* $E_0(\Lambda)\mathcal{H}$ *and* $E(\Lambda)\mathcal{H}$, *respectively, and the asymptotic completeness* (4.39) *holds.*

4.4. The time-dependent picture. Let us, finally, show that $W^\pm(H, H_0; J^\pm)$ coincide with the wave operators constructed in subsection 4.1. We proceed from the representation

$$(J^\pm e^{-iH_0t}f)(x) = (2\pi)^{-d/2} \int_{\mathbb{R}^d} e^{i\varphi^\pm(x,\xi)-i|\xi|^2t}\zeta^\pm(x,\xi)\psi(|\xi|^2)\hat{f}(\xi)d\xi, \quad f \in \mathcal{S}(\mathbb{R}^d). \tag{4.40}$$

Stationary points $\xi^\pm(x,t)$ of the phase function are determined by the equation

$$x + (\nabla_\xi\Phi^\pm)(x, \xi^\pm(x,t)) - 2\xi^\pm(x,t)t = 0. \tag{4.41}$$

Using estimate (4.27) on $\nabla_\xi\Phi^\pm$ we see that for large $|t|$ equation (4.41) has a unique solution $\xi^\pm(x,t)$ and

$$\xi^\pm(x,t) = (2t)^{-1}x + (2t)^{-1}(\nabla_\xi\Phi^\pm)(x,(2t)^{-1}x) + O(|t|^{-2\rho}). \tag{4.42}$$

Applying the stationary phase method to integral (4.40) and setting

$$\Xi(x,t) = \varphi^\pm(x, \xi^\pm(x,t)) - |\xi^\pm(x,t)|^2t, \quad \pm t > 0, \tag{4.43}$$

we find that (for any choice of the function $\zeta^\pm$ satisfying the assumptions above)

$$(J^\pm e^{-iH_0t}f)(x) = e^{i\Xi(x,t)}(2it)^{-d/2}\psi(|\xi^\pm(x,t)|^2)\hat{f}(\xi^\pm(x,t)), \quad t \to \pm\infty, \tag{4.44}$$

up to a term which tends to zero in $L_2(\mathbb{R}^d)$. Moreover, according to (4.42) we can replace here $\xi^\pm(x,t)$ by $(2t)^{-1}x$. Formula (4.43) can also be simplified if we use equations (4.41), (4.42) and neglect terms which tend to zero (compared to $|x/t|$) as $t \to \pm\infty$. This leads to the expression (4.7) for $\Xi(x,t)$. Thus it follows from (4.44) that

$$\lim_{t\to\pm\infty} ||J^\pm(\psi,\zeta^\pm)e^{-iH_0t}f - U_0(t)\psi(H_0)f|| = 0, \quad \forall f \in L_2(\mathbb{R}^d),$$

where $U_0(t)$ is operator (4.2). Consequently, the wave operators (3.5) and $W^\pm(H, H_0; J^\pm)$ exist simultaneously and coincide. This gives us

Theorem 4.8 *Suppose that condition* (4.1) *is fulfilled for* $\rho > 1/2$. *Let* $U_0(t)$ *be defined by equalities* (4.2), (4.7). *Then the wave operators* (3.5) *exist, are isometric and complete. Furthermore,*

$$\mathcal{W}^\pm(H, H_0)\psi(H_0) = W^\pm(H, H_0; J^\pm(\psi, \sigma^\pm, \eta)), \quad \psi \in C_0^\infty(\mathbb{R}_+). \tag{4.45}$$

4.5. The general case. In the case $\rho \leq 1/2$, proofs of the above results follow the same pattern. Of course, the phase $\Phi^\pm(x,\xi)$ can no longer be defined by (4.26), so one has to apply the method of successive approximations to equation (4.25) and to keep $[\rho^{-1}]$ iterations (cf. subsection 4.1). The main difficulty is that the standard PDO calculus fails for symbols from the classes $S^m_{\rho,1-\rho}$ if $\rho \leq 1/2$. Therefore to treat the general case, one has to modify (see [174]) PDO calculus taking into account an oscillating nature of symbols (4.28).

Finally, we mention that all constructions are preserved if a potential is a sum $V = V_L + V_S$ of a function V_L satisfying (4.1) and of a short-range term $V_S(x) = O(|x|^{-\rho_0})$, $\rho_0 > 1$, as $|x| \to \infty$. In particular, $\Phi^\pm$ is determined by equation (4.25) where V is replaced by V_L. The additional term $V_S J^\pm$ arising in (4.22) admits the factorization $V_S J^\pm = \langle x \rangle^{-l} B^\pm \langle x \rangle^{-l}$ with $l = \rho_0/2$ and a bounded operator $B^\pm$.

5. THE GENERALIZED FOURIER TRANSFORM

Another (compared to Section 4) approach to scattering on long-range potentials relies on a preliminary diagonalization of the Schrödinger operator $H = -\Delta + V$ via generalized Fourier transforms $F^{\pm}$. The operators $F^{\pm}$ are constructed in terms of spatial asymptotics of integral kernel (the Green function) of the resolvent $R(z) = (H - z)^{-1}$ for $z = \lambda \pm i0$. In this section we assume that $V = V_S + V_L$ where a short-range term V_S satisfies (1.15) for $\rho = \rho_s > 1$ and a long-range term V_L satisfies estimate (4.1) for some $\rho > 0$ and $|\kappa| \leq 3$.

5.1. Asymptotics of the Green function. To describe the asymptotics of functions $(R(\lambda \pm i0)f)(x)$ as $|x| \to \infty$, we need, first of all, to find an (approximate) solution $\phi(x, \lambda)$ of the eikonal equation

$$|\nabla \phi|^2 + V = \lambda$$

close to the function $\lambda^{1/2}|x|$ for large $|x|$. More precisely, we set

$$\phi(x, \lambda) = \lambda^{1/2}|x| - \Phi(x, \lambda) - (d-3)/4$$

and construct (see, for instance, [77], for details) a function $\Phi(x, \lambda)$ such that

$$2\lambda^{1/2}\partial\Phi(x, \lambda)/\partial|x| - |\nabla\Phi(x, \lambda)|^2 - V(x) = O(|x|^{-1-\varepsilon}), \quad \varepsilon > 0,$$

and

$$|(\partial_x^\alpha \partial_\lambda^\beta \Phi)(x, \lambda)| \leq C_{\alpha,\beta}(1 + |x|)^{1-|\alpha|-\varepsilon}. \tag{5.1}$$

For example, if condition (4.1) is satisfied for $\rho > 1/2$ we can set

$$\Phi(x, \lambda) = 2^{-1}\lambda^{-1/2}|x| \int_0^1 V_L(xt)dt. \tag{5.2}$$

Put also

$$w_{\pm}(x, \lambda) = |x|^{-(d-1)/2} e^{\pm i\phi(x,\lambda)}. \tag{5.3}$$

The main analytical result of this approach is formulated in the following

Theorem 5.1 *For any $f \in \mathbf{B}$*

$$(R(\lambda \pm i0)f)(x) = \pi^{1/2}\lambda^{-1/4}a^{\pm}(\pm\hat{x})w_{\pm}(x, \lambda) + \varepsilon_0(x), \tag{5.4}$$

$$(\partial_r R(\lambda \pm i0)f)(x) = \pm i\pi^{1/2}\lambda^{1/4}a^{\pm}(\pm\hat{x})w_{\pm}(x, \lambda) + \varepsilon_1(x) \tag{5.5}$$

where $a^{\pm} \in \mathfrak{N} = L_2(\mathbf{S}^{d-1})$ and $\varepsilon_0, \varepsilon_1 \in \mathbf{B}_0^*$.

If $f \in \mathcal{S} = \mathcal{S}(\mathbf{R}^d)$, then the estimate of the remainder in Theorem 5.1 can be improved. In this case

$$\int_{\mathbf{S}^{d-1}} |\varepsilon_j(r\omega)|^2 d\omega = o(r^{-d+1}), \quad r \to \infty, \quad j = 0, 1.$$

This result was first proven in [70] for $\rho > 1/2$ and for a suitably chosen sequence $r_n \to \infty$. These technical restrictions have been independently overcome in [77] and [136]. The extension of asymptotics (5.4), (5.5) to arbitrary $f \in \mathbf{B}$ is almost automatic and is discussed below. Of course, (5.4) reduces to (1.20) if $V = 0$.

The proof of Theorem 5.1 for $f \in \mathcal{S}$ relies on the limiting absorption principle (Theorem 4.2) and the following a priori estimate [72], [136] on the resolvent, known as the radiation estimate.

Theorem 5.2 *For any $l > 1/2$ and some $p = p(l, \rho, \rho_s) < 1/2$, the operator*

$$(\nabla \mp i\lambda^{1/2}\hat{x})R(\lambda \pm i0) : L_2^{(l)}(\mathbf{R}^d) \to L_2^{(-p)}(\mathbf{R}^d)$$

is bounded.

Compared to Theorem 4.3, this result contains also an information on the radial part of the function $(R(\lambda \pm i0)f)(x)$. As far as its angular part is concerned, Theorem 4.3 corresponds to the limit case of Theorem 5.2 for $l = p = 1/2$.

Note that the uniqueness theorem 3.11 is true under assumptions of this section.

Given (5.4), (5.5) for $f \in \mathcal{S}$ and $a^{\pm} = a^{\pm}(\lambda; f) \in \mathfrak{N}$, we define the mapping $\Gamma^{\pm}(\lambda) : \mathcal{S} \to \mathfrak{N}$ by the equality

$$(\Gamma^{\pm}(\lambda)f)(\hat{x}) = a^{\pm}(\hat{x}). \tag{5.6}$$

As follows from (1.20), in the free case $V = 0$, the operators $\Gamma^{+}(\lambda) = \Gamma^{-}(\lambda) = \Gamma_0(\lambda)$ where $\Gamma_0(\lambda)$ is defined by (1.21). In the long-range case formula (5.6) plays the role of representation (2.13) for short-range potentials.

Set $v_{\pm} = R(\lambda \pm i0)f$ for $f \in \mathcal{S}$ so that

$$(-\Delta + V - \lambda)v_{\pm} = f.$$

Applying the Green formula and using relations (5.4), (5.5), we obtain that (below $dS_r = r^{d-1}d\omega$)

$$\begin{aligned}\int_{|x|\le r} (v_{\pm}\bar{f} - f\bar{v}_{\pm})\, dx &= \int_{|x|\le r} (\bar{v}_{\pm}\Delta v_{\pm} - v_{\pm}\Delta\bar{v}_{\pm})\, dx = \int_{|x|=r} (\bar{v}_{\pm}\partial_r v_{\pm} - v_{\pm}\partial_r\bar{v}_{\pm})\, dS_r \\ &= \pm 2\pi i \int_{\mathbf{S}^{d-1}} |a^{\pm}(\omega)|^2\, d\omega + o(1)\end{aligned}$$

and hence, passing to the limit $r \to \infty$,

$$(R(\lambda + i0)f, f) - (R(\lambda - i0)f, f) = 2\pi i \int_{\mathbf{S}^{d-1}} |(\Gamma^{\pm}(\lambda)f)(\omega)|^2 d\omega. \tag{5.7}$$

Recall now that, by Theorem 4.2, the operator $R(\lambda \pm i0) : \mathbf{B} \to \mathbf{B}^*$ is bounded. Therefore relation (5.7) implies

Proposition 5.3 *The operator $\Gamma^{\pm}(\lambda) : \mathcal{S} \to \mathfrak{N}$ extends by continuity to a bounded operator from* $\mathbf{B}$ *into* $\mathfrak{N}$, *and hence the adjoint operator*

$$\Gamma^{\pm}(\lambda)^* : \mathfrak{N} \to \mathbf{B}^* \tag{5.8}$$

is also bounded.

In particular, (3.11) and Proposition 5.3 allow us to extend asymptotics (5.4), (5.5) to arbitrary $f \in \mathbf{B}$.

Our next aim is to express $\Gamma^{\pm}(\lambda)^* a$ for $a \in C^\infty(\mathbf{S}^{d-1})$ in terms of boundary values of the resolvent. Let $\eta \in C^\infty(\mathbf{R}^d)$ be such that $\eta(x) = 0$ in a neighbourhood of $x = 0$ and $\eta(x) = 1$ for large $|x|$. Obviously, the functions

$$u_\pm(x, \lambda) = \eta(x) a(\pm\hat{x}) w_\pm(x, \lambda) \tag{5.9}$$

belong to the space $\mathbf{B}^*$. Set also

$$g_\pm(\lambda) = (-\Delta + V - \lambda) u_\pm(\lambda). \tag{5.10}$$

Straightforward calculations show that

$$g_\pm(x, \lambda) = e^{\pm i\phi(x,\lambda)} b_\pm(x, \lambda) \quad \text{where} \quad \partial^\kappa b_\pm(x, \lambda) = O(|x|^{-(d+1)/2-\rho-|\kappa|}) \tag{5.11}$$

as $|x| \to \infty$. In particular, $g_\pm(\lambda) \in L_2^{(l)}(\mathbf{R}^d)$ for $l < 1/2 + \rho$. The following assertion can be deduced from Theorem 5.1.

Proposition 5.4 *Let $a \in C^\infty(\mathbf{S}^{d-1})$ and let $u_\pm$, $g_\pm$ be given by* (5.9), (5.10). *Then* (*as elements of* $\mathbf{B}^*$)

$$\pm 2i\pi^{1/2}\lambda^{1/4}\Gamma^{\pm}(\lambda)^* a = u_\pm(\lambda) - R(\lambda \mp i0) g_\pm(\lambda). \tag{5.12}$$

In particular, the operator-function $\Gamma^{\pm}(\lambda)^* : \mathfrak{N} \to L_2^{(-l)}(\mathbf{R}^d)$ is strongly continuous in $\lambda > 0$ for any $l > 1/2$.

Indeed, for arbitrary $f \in \mathbf{B}$, applying the Green formula to $u_\pm$ and $v_\pm = R(\lambda \pm i0)f$, we see that

$$\int_{|x|\le r} (u_\pm \overline{f} - g_\pm \overline{v}_\pm)\, dx = \int_{|x|=r} (\overline{v}_\pm \partial_r u_\pm - u_\pm \partial_r \overline{v}_\pm)\, dS_r.$$

The left-hand side here tends to $(u_\pm, f) - (g_\pm, R(\lambda \pm i0)f)$ as $r \to \infty$. According to Theorem 5.1, the right-hand side converges, at least by some sequence $r_n \to \infty$, to $\pm 2i\pi^{1/2}\lambda^{1/4}(a, \Gamma^{\pm}(\lambda)f)$. This yields (5.12).

It follows from (5.10), (5.12) that

$$(-\Delta + V - \lambda)\Gamma^{\pm}(\lambda)^* a = 0. \tag{5.13}$$

Moreover, (5.8) enables us to extend this result to all $a \in \mathfrak{N}$.

Corollary 5.5 *For any $a \in \mathfrak{N}$, the function $\Gamma^{\pm}(\lambda)^* a \in \mathsf{H}^2_{loc} \cap \mathbf{B}^*$ and* (5.13) *holds.*

Putting together Theorem 5.1 and Proposition 5.4, we see that

$$\left.\begin{aligned}(\Gamma^{\pm}(\lambda)^{*}a)(x) &= a_{+}^{\pm}(\hat{x})w_{+}(x,\lambda) - a_{-}^{\pm}(-\hat{x})w_{-}(x,\lambda) + \varepsilon_{0}(x),\\ (\partial_{r}\Gamma^{\pm}(\lambda)^{*}a)(x) &= i\lambda^{1/2}a_{+}^{\pm}(\hat{x})w_{+}(x,\lambda) + i\lambda^{1/2}a_{-}^{\pm}(-\hat{x})w_{-}(x,\lambda) + \varepsilon_{1}(x)\end{aligned}\right\} \tag{5.14}$$

for some functions $a_{+}^{\pm}, a_{-}^{\pm} \in \mathfrak{N}$ and $\varepsilon_0, \varepsilon_1 \in \mathbf{B}_0^*$. Moreover,

$$a_{+}^{+}(\hat{x}) = a_{-}^{-}(\hat{x}) = -i2^{-1}\pi^{-1/2}\lambda^{-1/4}a(\hat{x}). \tag{5.15}$$

Using the Green formula it is easy to deduce from equation (5.13) that

$$||a_{+}^{\pm}||_{\mathfrak{N}} = ||a_{-}^{\pm}||_{\mathfrak{N}}.$$

Taking into account (5.8), we can now extend (5.14) to all $a \in \mathfrak{N}$. Asymptotics (5.14) implies that

$$\lim_{r\to\infty} r^{-1}\int_{|x|\le r} |(\Gamma^{\pm}(\lambda)^{*}a)(x)|^{2}dx = (2\pi)^{-1}\lambda^{-1/2}||a||_{\mathfrak{N}}^{2}.$$

Combined with (5.8) this gives the two-sided estimate

$$(2\pi)^{-1/2}\lambda^{-1/4}||a||_{\mathfrak{N}} \le ||\Gamma^{\pm}(\lambda)^{*}a|| \le C||a||_{\mathfrak{N}}. \tag{5.16}$$

In particular, (5.16) shows that Proposition 5.3 and hence estimate (3.11) are optimal.

The results discussed above permit (see [55]) to generalize Theorem 3.10 to long-range potentials. The only difference is that the usual spherical waves are replaced by modified waves (5.3). Actually, (5.14), (5.15) mean that for any $a_+ \in \mathfrak{N}$ (or $a_- \in \mathfrak{N}$) there exists a (unique) solution u of the Schrödinger equation (1.14) with asymptotics

$$\left.\begin{aligned}u(x) &= a_{+}(\hat{x})w_{+}(x) - a_{-}(-\hat{x})w_{-}(x) + \varepsilon_{0}(x),\\ (\partial_{r}u)(x) &= i\lambda^{1/2}a_{+}(\hat{x})w_{+}(x) + i\lambda^{1/2}a_{-}(-\hat{x})w_{-}(x) + \varepsilon_{1}(x),\end{aligned}\right\} \tag{5.17}$$

where $\varepsilon_0, \varepsilon_1 \in \mathbf{B}_0^*$. It can also be proven that every solution $u \in \mathsf{H}^2_{loc} \cap \mathbf{B}^*$ of equation (1.14) has asymptotics (5.17) with $a_\pm \in \mathfrak{N}$. Coefficients $a_\pm$ are related by the scattering matrix $S(\lambda)$ corresponding to modified wave operators (3.5): $a_+ = S(\lambda)a_-$. Construction of this subsection gives the following representations for the scattering matrix:

$$S(\lambda)a = \pi^{1/2}\lambda^{-1/4}\Gamma_{+}(\lambda)(-\Delta + V - \lambda)u_{-}(\lambda),$$

$$S^{*}(\lambda)a = \pi^{1/2}\lambda^{-1/4}\Gamma_{-}(\lambda)(-\Delta + V - \lambda)u_{+}(\lambda),$$

where $u_\pm(\lambda)$ are related to a by formula (5.9). These representations were first obtained in [137] (see also [55]).

5.2. Connection to wave operators. The results of the previous subsection allows one also to derive an expansion theorem in eigenfunctions of the operator H

in a form similar to that of subsection 2.3. As before, $\mathfrak{N} = L_2(\mathbb{S}^{d-1})$, $\mathfrak{H} = L^2(\mathbb{R}_+;\mathfrak{N})$ is the Hilbert space of $\mathfrak{N}$-valued square integrable functions on $\mathbb{R}_+$ with the Lebesgue measure. Define again the generalized Fourier transform $F^\pm : \mathbf{B} \to \mathfrak{H}$ by formula (2.14) but now $\Gamma^\pm(\lambda)$ is the operator (5.6). Integrating (5.7) over $\mathbb{R}_+$ and using (1.2), (1.3), we find that (2.15) holds for any $f \in \mathbf{B}$. Hence the operator $F^\pm$ extends by continuity to a bounded operator on the entire space $\mathcal{H}$. This operator is isometric on the absolutely continuous subspace $\mathcal{H}^{(ac)} = P\mathcal{H}$ of the operator H and vanishes on its orthogonal complement, i.e., equality (2.16) is satisfied. The intertwining property (2.17) is an easy consequence of equation (5.13). Finally, the identity (2.19) can be deduced from the lower bound (5.16) which implies that the kernel of the operator $\Gamma^\pm(\lambda)^*$ is trivial. Let us summarize these results.

Theorem 5.6 *For every $f \in \mathbf{B}$ the function $\Gamma^\pm(\lambda)f$ belongs to the space $\mathfrak{H}$. The operator $F^\pm : \mathcal{H} \to \mathfrak{H}$ defined by* (2.14) *extends by continuity to a bounded operator on the entire space $\mathcal{H}$ and satisfies relations* (2.16), (2.17) *and* (2.19).

Theorem 5.6 can be used for the proof of the existence of the modified time-dependent wave operators (3.5) and of their completeness. Indeed, choose $\tilde{f}(\lambda) = v(\lambda)a$ where $a \in C^\infty(\mathbb{S}^{d-1})$ and $v \in C_0^\infty(\mathbb{R}_+)$. Linear combinations of such $\tilde{f}$ are dense in $\mathfrak{H}$. It follows from the intertwining property (2.17) that

$$e^{-iHt}f = \int_0^\infty e^{-i\lambda t} v(\lambda)\Gamma^\pm(\lambda)^* a \, d\lambda, \tag{5.18}$$

where the function $\Gamma^\pm(\lambda)^*a$ satisfies (5.12). Following [71], we shall check that the second term in the right-hand side of (5.12) is negligible.

Lemma 5.7 *Let $g_\pm(\lambda)$ be defined by* (5.10). *Then the norm of the function*

$$G_\pm(t) = \int_0^\infty e^{-i\lambda t} v(\lambda) R(\lambda \mp i0) g_\pm(\lambda) d\lambda$$

tends to zero as $t \to \pm\infty$.

Consider, for example, the upper sign and set

$$\mathcal{G}_+(t) = \int_0^\infty e^{-i\lambda t} v(\lambda) g_+(\lambda) d\lambda. \tag{5.19}$$

It is easy to deduce from the representation

$$R(\lambda - i\varepsilon) = -ie^{-i(H-\lambda+i\varepsilon)t} \int_t^\infty e^{i(H-\lambda+i\varepsilon)s} ds, \quad \varepsilon > 0, \quad \forall t \in \mathbb{R},$$

that

$$||G_+(t)|| \leq \int_t^\infty ||\mathcal{G}_+(s)|| ds. \tag{5.20}$$

Let us now use representation (5.11) for $g_+(\lambda)$. According to (5.1) the function $-\lambda t + \phi(x, \lambda)$ does not have stationary points if $|x| \leq \alpha|t|$ or $|x| \geq \beta|t|$ where α is small and β is large enough. Therefore, integrating by parts in (5.19), we see that, for arbitrary N,

$$|\mathcal{G}_+(x,t)| \leq C_N(1 + |t|)^{-N} \quad \text{or} \quad |\mathcal{G}_+(x,t)| \leq C_N(1 + |x|)^{-N}$$

if $|x| \leq \alpha|t|$ or $|x| \geq \beta|t|$, respectively. In the region $\alpha|t| \leq |x| \leq \beta|t|$, the stationary phase method allows us to gain an additional factor $t^{-1/2}$ compared to (5.11). This yields the estimate

$$|\mathcal{G}_+(x,t)| \leq C(1+|x|)^{-(d+1)/2-\rho}t^{-1/2}$$

and consequently $||\mathcal{G}_+(t)|| = O(t^{-1-\rho})$. Now it follows from (5.20) that $||G_+(t)|| \to 0$ as $t \to \infty$.

Thus asymptotics of function (5.18) as $t \to \pm\infty$ is determined by the term $u_\pm(\lambda)$ in the right-hand side of (5.12). Taking into account its definition (5.9), we obtain that

$$(e^{-iHt}f)(x) = \mp i2^{-1}\pi^{-1/2}|x|^{-(d-1)/2}\eta(x)a(\pm\hat{x})I_\pm(x,t) + \epsilon_\pm(x,t), \tag{5.21}$$

where

$$I_\pm(x,t) = \int_0^\infty e^{-i\lambda t \pm i\phi(x,\lambda)}v(\lambda)\lambda^{-1/4}d\lambda \tag{5.22}$$

and $||\epsilon_\pm(\cdot,t)|| = o(1)$ as $t \to \pm\infty$. The asymptotics of integral (5.22) can be found by the stationary phase method. The regions $|x| \leq \alpha|t|$ and $|x| \geq \beta|t|$, for sufficiently small α and large β, are again negligible. In the layer $\alpha|t| \leq |x| \leq \beta|t|$ the asymptotics of (5.22) is determined by stationary points $\lambda_s = \lambda_s(x,t)$ satisfying the equation

$$\lambda_s^{1/2} = (2|t|)^{-1}|x| - |t|^{-1}\Phi_\lambda(x,\lambda_s)\lambda_s^{1/2}. \tag{5.23}$$

Therefore

$$I_\pm(x,t) = (2\pi)^{1/2}|x|^{1/2}|t|^{-1}\exp\big(i\Xi(x,t) \mp \pi i(d-2)/2\big)v(4^{-1}|t|^{-2}|x|^2) + o(|t|^{-1/2}), \tag{5.24}$$

where

$$\Xi(x,t) = -\lambda_s t \pm \phi(x,\lambda_s(x,t)) \pm \pi(d-3)/4. \tag{5.25}$$

Using equation (5.23) we can check that function (5.25) coincides with the function $\Xi(x,t)$ constructed in subsection 4.1. In particular, in the case $\rho > 1/2$ the function $\Phi(x,\lambda)$ can be defined by (5.2); then $\Xi(x,t)$ is given by (4.7).

Let the modified free evolution $U_0(t)$ be defined by equality (4.2). Recall that the operator F_0 was introduced by equations (1.21), (2.12). It follows from unitarity of this operator that

$$(\mathcal{F}_0F_0^*\tilde{f})(\xi) = 2^{1/2}|\xi|^{-(d-2)/2}v(|\xi|^2)a(\hat{\xi}) \quad \text{if} \quad \tilde{f}(\lambda) = v(\lambda)a.$$

Thus comparing (5.21) and (5.24), we obtain that

$$\lim_{t\to\pm\infty} ||e^{-iHt}f - U_0(t)F_0^*\tilde{f}|| = 0, \quad f = (F^\pm)^*\tilde{f}.$$

It follows that the wave operators (3.5) exist and

$$\mathcal{W}^\pm = (F^\pm)^*F_0. \tag{5.26}$$

Equality (2.16) is equivalent to their completeness. Let us formulate the result obtained.

Theorem 5.8 *Assume that* $V = V_S + V_L$ *where a short-range term* V_S *satisfies* (1.15) *for* $\rho = \rho_s > 1$ *and a long-range term* V_L *satisfies estimate* (4.1) *for some* $\rho > 0$ *and* $|\kappa| \leq 3$. *Let* $U_0(t)$ *be operator* (4.2) *with the function* $\Xi(x,t)$ *satisfying* (4.4) - (4.6). *Then the wave operators* (3.5) *exist, are isometric and complete, and equality* (5.26) *holds.*

Finally, we note that another (compared to (5.6)) representation of the operators $\Gamma^{\pm}(\lambda)$ and hence of $F^{\pm}$ may be found in subsection 7.2. This representation generalizes (2.13) for the short-range case.

6. LONG-RANGE MATRIX POTENTIALS

Here we consider the operator $H = -\Delta + V(x)$ with a Hermitian $(n \times n)$-matrix-function $V(x)$ in the space $\mathcal{H} = L_2(\mathbf{R}^d; \mathbf{C}^n)$.

6.1. Modified wave operators. There is no difference compared to the scalar case $n = 1$ for short-range potentials $V(x)$ satisfying condition (1.15) for some $\rho > 1$. In particular, Theorems 2.4 and 2.5 hold.

For long-range matrix-valued potentials $V(x)$, Theorems 4.2 and 4.3, that is the limiting absorption principle and the radiation estimates, are also true. In particular, the operator H does not have the singular continuous spectrum. The proofs of all these results are identical to the scalar case.

The difference between the scalar and vector cases lies in a construction of a modified free evolution. In the scalar case this construction is based on solutions of eikonal equations (4.3) or (4.25). This approach meets with some difficulties in the vector case (for systems of equations). To single out problems specific for the vector case we suppose for simplicity that $d = 1$.

Let us start with a generalization of the construction of subsection 4.1. Set

$$\Omega(x,t) = \exp(i(4t)^{-1}x^2)(2it)^{-1/2}.$$

According to (4.2) and (4.4) it is natural to seek a modified free evolution in the form

$$(U_0(t)f)(x) = \Omega(x,t)Y(t,x/t)\hat{f}(x/(2t)), \quad \hat{f} = \mathcal{F}f, \tag{6.1}$$

where $Y(t,x/t)$ is a family of unitary operators. It should be chosen in such a way that condition (3.6) is satisfied for functions $\hat{f} \in C_0^\infty(\mathbf{R} \setminus \{0\}; \mathbf{C}^n)$. Then wave operators (3.5) exist. Thus we need to construct the family $Y(t,s)$ for $s \in \mathbf{R} \setminus \{0\}$ and large $|t|$. A simple calculation shows that

$$\partial_t U_0(t)f = \Omega_t Y\hat{f} + \Omega(Y_t\hat{f} - t^{-2}xY_s\hat{f} - 2^{-1}t^{-2}xY\hat{f}') \tag{6.2}$$

and

$$\partial_x^2 U_0(t)f = \Omega_{xx}Y\hat{f} + \Omega t^{-2}(ixY_s\hat{f} + i2^{-1}xY\hat{f}' + Y_{ss}\hat{f} + Y_s\hat{f}' + 4^{-1}Y\hat{f}''). \tag{6.3}$$

Comparing (6.2), (6.3) and taking into account the equation $i\Omega_t + \Omega_{xx} = 0$, we see that

$$i\partial_t U_0(t)f + \partial_x^2 U_0(t)f = \Omega[(iY_t + t^{-2}Y_{ss})\hat{f} + t^{-2}Y_s\hat{f}' + 4^{-1}t^{-2}Y\hat{f}'')].$$

Here the term containing $\hat{f}''$ is $O(t^{-2})$, and hence it can be neglected. Therefore condition (3.6) is satisfied if

$$iY_t(t,s) - V(st)Y(t,s) + t^{-2}Y_{ss}(t,s) = O(|t|^{-1-\varepsilon}) \tag{6.4}$$

and

$$Y_s(t,s) = O(|t|^{1-\varepsilon}) \tag{6.5}$$

for some $\varepsilon > 0$ uniformly in s from a compact subset $K \subset \mathbb{R}\setminus\{0\}$. In the matrix case equation (6.4) (with zero in the right-hand side) plays the role of eikonal equation (4.3).

Suppose that condition (4.1) on matrix-valued potential $V(x)$ is fulfilled for all κ. The cases $\rho > 1/2$ and $\rho \in (0, 1/2]$ seem now to be essentially different. In the first case we may define Y as a solution of the equation

$$iY_t(t,s) = V(st)Y(t,s) \tag{6.6}$$

which becomes unique if some initial condition, for example $Y(0,s) = I$, is added. Of course, $Y(t,s)$ is unitary. Clearly, (6.1) coincides (see (4.7)) with (4.2) if $n = 1$.

It is easy to check that for any p

$$(\partial_s^p Y)(t,s) = O(|t|^{p(1-\rho)}). \tag{6.7}$$

Indeed, let us differentiate equation (6.6) in the parameter s and set $Y_s = YZ$. Then

$$iZ_t(t,s) = tY^*(t,s)V'(st)Y(t,s)$$

and hence $Z(t,s) = O(|t|^{1-\rho})$ which is equivalent to (6.7) for $p = 1$. This implies of course (6.5). For $p \geq 2$, condition (6.7) can be verified quite similarly. In particular, (6.7) for $p = 2$ shows that condition (6.4) is satisfied if $2\rho > 1$. Moreover, it can be easily deduced from equation (6.6) that relation (4.8) also holds. Thus we obtain

Proposition 6.1 *Suppose that condition* (4.1) *is fulfilled for $\rho > 1/2$ and all κ. Let $U_0(t)$ be defined by equations* (6.1), (6.6). *Then the wave operators* (3.5) *exist, are isometric and the intertwining property $H\mathcal{W}^{\pm} = \mathcal{W}^{\pm}H_0$ holds.*

The completeness of modified wave operators was verified (for $\rho > 1/2$ and arbitrary d) in [41], where operators $U_0(t)$ were, however, defined in the momentum representation.

The solution $Y(t,s)$ of equation (6.6) corresponds to the first approximation (4.7) to a solution of the eikonal equation (4.3). If $\rho \leq 1/2$, then in the scalar case one has to take higher approximations to a solution of this equation (see subsection 4.1). Apparently, in the matrix case these additive approximations of the phase function correspond to multiplicative approximations of the unitary family $Y(t,s)$. Namely, as the second approximation, let us seek an operator-function $Y(t,s)$ satisfying (6.4) in the form

$$Y(t,s) = Y^{(0)}(t,s)Y^{(1)}(t,s) \tag{6.8}$$

where $Y^{(0)}(t,s)$ is the solution of equation (6.6). Plugging (6.8) into (6.4), we see that $Y^{(1)}(t,s)$ should be defined as a solution of the equation

$$iY_t^{(1)}(t,s) = -t^{-2}\operatorname{Re}(Y^{(0)}(t,s)^* Y_{ss}^{(0)}(t,s))\, Y^{(1)}(t,s). \tag{6.9}$$

Then (6.4) is equivalent to the condition

$$\operatorname{Im}((Y^{(0)})^* Y_{ss}^{(0)})Y^{(1)} + 2(Y^{(0)})^* Y_s^{(0)} Y_s^{(1)} + Y_{ss}^{(1)} = O(|t|^{1-\varepsilon}),$$

which does not look obvious. Note that we have taken the symmetric part of the operator $(Y^{(0)})^* Y_{ss}^{(0)}$ in (6.9) in order not to lose the unitarity of the operators $Y^{(1)}$.

6.2. The existence of stationary wave operators. Let us now turn to the construction of subsection 4.3 which allowed us to check both the existence and completeness of wave operators $W^{\pm}(H, H_0; J)$. We emphasize that in the one-dimensional case it suffices to introduce only one identification $J = J^{\pm}$. It is defined by the equality

$$(Jf)(x) = (2\pi)^{-1/2} \int_{-\infty}^{\infty} e^{ix\xi} \mathbf{u}(x,\xi)\psi(\xi^2)\hat{f}(\xi)d\xi \tag{6.10}$$

(cf. (4.21), (4.28)), where $\mathbf{u}(x,\xi)$ is a unitary operator-valued function and $\psi \in C_0^\infty(\mathbb{R}_+)$. Clearly, $T = HJ - JH_0$ is also a PDO with symbol

$$t(x,\xi) = (-2i\xi\, \mathbf{u}_x(x,\xi) - \mathbf{u}_{xx}(x,\xi) + V(x)\mathbf{u}(x,\xi))\psi(\xi^2). \tag{6.11}$$

Suppose that condition (4.1) is fulfilled for some $\rho > 0$ and all κ. Let $\mathbf{v}$ be determined (for sufficiently large $|x|$) by the equation

$$2\xi\, \mathbf{v}(x,\xi) = V(x) + \mathbf{v}(x,\xi)^2,$$

that is

$$\mathbf{v}(x,\xi) = \xi I - (\xi^2 I - V(x))^{1/2}; \tag{6.12}$$

in a compact region of x the function $\mathbf{v}(x,\xi) = \mathbf{v}(x,\xi)^*$ may be arbitrary. Clearly,

$$\mathbf{v}\psi \in S_{1,0}^{-\rho}. \tag{6.13}$$

We define $\mathbf{u}(x,\xi)$ as a solution of the equation

$$i\mathbf{u}_x(x,\xi) = \mathbf{v}(x,\xi)\mathbf{u}(x,\xi) \tag{6.14}$$

(complemented, for example, by the condition $\mathbf{u}(0,\xi) = I$). Then symbol (6.11) equals

$$t(x,\xi) = i\mathbf{v}_x(x,\xi)\mathbf{u}(x,\xi)\psi(\xi^2). \tag{6.15}$$

In the previous subsection we have met a problem of estimating derivatives of $Y(t,s)$ in s for large $|t|$. A similar problem arises by the study of behaviour of derivatives of $\mathbf{u}(x,\xi)$ in the variable ξ for large $|x|$. Indeed, let us differentiate equation (6.14) in the parameter ξ and set $\mathbf{u}_\xi = \mathbf{u}\mathbf{z}$. Then

$$i\mathbf{z}_x(x,\xi) = \mathbf{u}^*(x,\xi)\mathbf{v}_\xi(x,\xi)\mathbf{u}(x,\xi), \tag{6.16}$$

so $\mathbf{z}(x,\xi) = O(|x|^{1-\rho})$ and hence $\mathbf{u}_\xi(x,\xi) = O(|x|^{1-\rho})$. Repeating this procedure, one can check that

$$\partial_\xi^p \mathbf{u}(x,\xi) = O(|x|^{p(1-\rho)}). \tag{6.17}$$

Estimate (6.17) suffices to justify the usual formula of commutation of PDO T with symbol (6.15) and of the operator $\langle x\rangle^{-l}$. Hence condition (6.13) ensures that T can be factorized as

$$T = \langle x\rangle^{-l} B \langle x\rangle^{-l}, \quad l = (1+\rho)/2,$$

with a bounded operator B. Now Proposition 1.18 implies

Proposition 6.2 *Let condition* (4.1) *be fulfilled for some* $\rho > 0$, *and let* $\mathbf{v}(x,\xi)$ *be matrix-function* (6.12). *Define the PDO* J *by formula* (6.10) *where* $\mathbf{u}(x,\xi)$ *satisfies equation* (6.14). *Then the wave operators*

$$W^\pm(H,H_0;J), \quad W^\pm(H_0,H;J^*) \tag{6.18}$$

exist.

Note that expression (6.12) for $\mathbf{v}(x,\xi)$ can be replaced by a finite number (which depends on ρ) of terms in the Taylor series

$$\mathbf{v}(x,\xi) = (2\xi)^{-1}V(x) + (2\xi)^{-3}V(x)^2 + \dots.$$

This does not change wave operators (6.18).

6.3. The completeness and isometricity of wave operators. Formally, the symbol of PDO JJ^* admits the standard asymptotic expansion

$$\sum_{p=0}^{\infty}(p!)^{-1} i^{-p}\partial_\xi^p(\mathbf{j}(x,\xi)\partial_x^p \mathbf{j}^*(x,\xi)), \tag{6.19}$$

where $\mathbf{j}(x,\xi) = \mathbf{u}(x,\xi)\psi(\xi)$ is the symbol of PDO J. Such expansions were justified (this is non-trivial for $\rho \le 1/2$ only) in [174] for scalar PDO with oscillating symbols. In the matrix case the proof is the same and basically reduces to verification of inclusions

$$\partial_\xi^p(\mathbf{j}(x,\xi)\partial_x^p \mathbf{j}^*(x,\xi)) \in S_{1,0}^{-p\rho}. \tag{6.20}$$

Then (6.19) is really an asymptotic series. To check (6.20) we use equation (6.14). For example, for $p=1$ it suffices to remark that $\mathbf{u}(x,\xi)\partial_x\mathbf{u}^*(x,\xi) = i\mathbf{v}(x,\xi)$ and $\mathbf{v}$ satisfies (6.13). Since $\psi^2(H) - \psi^2(H_0) \in \mathfrak{S}_\infty$, expansion (6.19) implies that

$$JJ^* - \psi^2(H) \in \mathfrak{S}_\infty.$$

Supposing that $\psi(\lambda) = 1$ on some compact interval $\Lambda \subset (0,\infty)$, we see that the operators $W^{\pm}(H_0, H; J^*)$ are isometric on $E(\Lambda)\mathcal{H}$.

It remains to check isometricity of the operators $W^{\pm}(H, H_0; J)$. Similarly to (6.19), the symbol of PDO J^*J admits the formal asymptotic expansion

$$\sum_{p=0}^{\infty}(p!)^{-1}i^{-p}\partial_x^p(\mathbf{j}^*(x,\xi)\partial_\xi^p\mathbf{j}(x,\xi)), \tag{6.21}$$

which is however much less convenient than (6.19). Indeed, consider, for example, the term corresponding to $p = 1$. According to equation (6.16) it equals

$$-i\partial_x(\mathbf{u}^*(x,\xi)\partial_\xi\mathbf{u}(x,\xi))\psi^2(\xi^2) = -\mathbf{u}^*(x,\xi)\mathbf{v}_\xi(x,\xi)\mathbf{u}(x,\xi)\psi^2(\xi^2) =: \mathbf{r}(x,\xi). \tag{6.22}$$

By (6.13), the PDO R with symbol $\mathbf{r}(x,\xi)$ belongs to the Hilbert-Schmidt class if $2\rho > 1$. Probably terms in (6.21) corresponding to $p > 1$ can be treated similarly, so

$$J^*J - \psi^2(H_0) \in \mathfrak{S}_\infty. \tag{6.23}$$

We shall give the precise formulation of the results discussed above although the proof of (6.23) was not quite complete.

Proposition 6.3 *Under the assumptions of Proposition* 6.2 *the wave operators* $W^{\pm}(H, H_0; J)$ *exist and are complete, that is equality* (4.39) *where* $J^{\pm} = J$ *holds. Moreover, in the case* $\rho > 1/2$ *the wave operators* $W^{\pm}(H, H_0; J)$ *are isometric.*

To verify that $R \in \mathfrak{S}_\infty$ in the case $\rho \in (0, 1/2]$, one needs to estimate the derivatives of symbol (6.22) in the variable ξ. However inequality (6.17) is not sufficient to control these derivatives. Nevertheless, estimates are getting better if one takes powers of R. For example, the symbol of the PDO R^2 equals

$$\mathbf{r}^2 - i\partial_\xi(\mathbf{r}\partial_x\mathbf{r}) + \ldots = (\mathbf{u}^*\mathbf{v}_\xi^2\mathbf{u} - i\partial_\xi(\mathbf{u}^*\mathbf{v}_\xi\mathbf{v}_{x\xi}\mathbf{u})\psi^4 + \ldots,$$

where by calculation of $\mathbf{r}\partial_x\mathbf{r}$ we have used equation (6.14) and the identity $\mathbf{v}_\xi\mathbf{v} = \mathbf{v}\mathbf{v}_\xi$. It follows from (6.13) and (6.17) that

$$\mathbf{u}^*\mathbf{v}_\xi^2\mathbf{u} = O(|x|^{-2\rho}), \quad \partial_\xi(\mathbf{u}^*\mathbf{v}_\xi\mathbf{v}_{x\xi}\mathbf{u}) = O(|x|^{-3\rho}).$$

Thus we can expect that the operator R^2 belongs the Hilbert-Schmidt class and hence R is compact if $\rho > 1/4$. Considering higher powers of R, one can hope to check that $R \in \mathfrak{S}_\infty$ for any $\rho > 0$. Therefore it is sufficiently plausible that inclusion (6.23) holds and $W^{\pm}(H, H_0; J)$ are isometric for all $\rho > 0$.

In the case $\rho > 1/2$ it is natural to expect (cf. subsection 4.4) that the wave operators $W^{\pm}(H, H_0; J)$ and $\mathcal{W}^{\pm}(H, H_0)$ are related by equality (4.45). This gives an efficient description of the asymptotics as $t \to \pm\infty$ of the function $\exp(-iHt)f$ for any $f \in \mathcal{H}^{(ac)}$ in terms of the family (6.1).

For an arbitrary $\rho > 0$, Proposition 6.3 guarantees that

$$\lim_{t\to\pm\infty} ||\exp(-iHt)f - J\exp(-iH_0t)f_\pm|| = 0, \quad \forall f \in \mathcal{H}^{(ac)}.$$

An explicit description of the asymptotics of the function $J\exp(-iH_0t)f_\pm$ can be based on the stationary phase method. This requires however the asymptotics as $|x| \to \infty$ of solutions of the system of n equations (6.14). Under the only assumption (4.1) this looks like an open problem. We emphasize that Ansatz (6.1) is actually very general. Nevertheless, surprisingly, it seems not to work in the case $\rho \le 1/2$.

Thus we finish with

Problem 6.4 *Develop a coherent scattering theory for the Schrödinger operator with a long-range matrix-valued potential.*

Part II

The scattering matrix

The scattering matrix $S(\lambda)$ for the Schrödinger operator $H = H_0 + V$ is a unitary operator in the space $\mathfrak{N} = L_2(\mathbb{S}^{d-1})$. Regarded as an integral operator, it has a smooth kernel off the diagonal $\omega = \omega'$. Its diagonal singularity is radically different for short- and long-range perturbations V. In the first case the leading singularity is the Dirac-function. In the second the singularity is essentially more strong and has an oscillating character. The nature of the diagonal singularity determines the difference in spectral properties of the scattering matrix for short- and long-range interactions. An alternative, and often more convenient, point of view is to regard $S(\lambda)$ as a pseudo-differential operator (PDO) and to analyze the asymptotics of its symbol at infinity.

7. A STATIONARY REPRESENTATION

Our study of the scattering matrix $S(\lambda)$ relies on its representation in terms of the resolvent. As we have seen in Section 4, in the long-range case the stationary approach requires an introduction of non-trivial identifications $J^{\pm}$ different for $t \to \pm\infty$. In this section we derive a stationary formula for $S(\lambda)$ in the general abstract framework.

7.1. Definition of the scattering matrix. Recall that the scattering operator $\mathbf{S}$ defined by formula (1.12) commutes with H_0 and hence reduces to multiplication by the operator-function (the scattering matrix)

$$S(\lambda) = S(\lambda; H, H_0; J^+, J^-) \tag{7.1}$$

in a spectral representation of the operator H_0. To be more precise, suppose that the spectrum of H_0 has constant multiplicity κ on an interval Λ of the spectral axis, and let $\mathfrak{N}$ be an auxiliary Hilbert space such that $\dim \mathfrak{N} = \kappa$. By the spectral theorem, there exists a unitary operator

$$F_0 : E_0(\Lambda)\mathcal{H}_0 \to L_2(\Lambda; \mathfrak{N}) \tag{7.2}$$

such that

$$(F_0H_0f)(\lambda) = \lambda(F_0f)(\lambda). \tag{7.3}$$

It follows that

$$||(F_0f)(\lambda)||_{\mathfrak{N}}^2 = d(E_0f, f)/d\lambda$$

for almost all $\lambda \in \Lambda$. We use also a formal notation $\Gamma_0(\lambda)$ defined by (2.12); then

$$||\Gamma_0(\lambda)f||_{\mathfrak{N}}^2 = d(E_0f, f)/d\lambda. \tag{7.4}$$

In applications the operator $\Gamma_0(\lambda)$ is well-defined and continuous in λ on a suitable dense in $\mathcal{H}_0$ set of elements f.

Since $\mathbf{S}H_0 = H_0\mathbf{S}$,

$$(\mathbf{S}F_0f)(\lambda) = S(\lambda)(F_0f)(\lambda), \tag{7.5}$$

where the scattering matrix

$$S(\lambda) : \mathfrak{N} \to \mathfrak{N}.$$

It is unitary for almost all $\lambda \in \Lambda$ if the scattering operator $\mathbf{S}$ is unitary on $E_0(\Lambda)\mathcal{H}_0$. Of course, the scattering matrix $S(\lambda)$ depends on the operator F_0 (on the choice of the spectral representation of H_0) and hence, in the abstract framework, it is defined up to a unitary equivalence only. In particular, the spectrum of $S(\lambda)$ does not depend on F_0. Obviously, the scattering matrix $S(\lambda)$ is not changed for $\lambda \in \Lambda$ if $J^\pm$ in (7.1) are replaced by $J^\pm E_0(\Lambda)$.

All our considerations rely on stationary formulas for operator (7.1). To obtain them, we have to start with a generalization of stationary representation (2.22) to wave operators (1.8).

7.2. Wave operators in a couple of spaces. Here we extend the approach of subsection 2.3 to wave operators $W^\pm(H, H_0; J)$ with a non-trivial identification $J \neq I$. Clearly, the role of (1.13) is played by the resolvent identity

$$R(z)J = JR_0(z) - R(z)TR_0(z), \quad T = HJ - JH_0, \quad \operatorname{Im} z \neq 0.$$

Instead of (2.20) and (2.21) we have more general identities

$$2\varepsilon \int_0^\infty e^{-2\varepsilon t}(J\exp(\pm itH_0)f_0, \exp(\pm itH)f)dt$$
$$= \pi^{-1}\varepsilon \int_{-\infty}^\infty (JR_0(\lambda \pm i\varepsilon)f_0, R(\lambda \pm i\varepsilon)f)d\lambda, \tag{7.6}$$

and

$$\pi^{-1}\varepsilon(JR_0(\lambda \pm i\varepsilon)f_0, R(\lambda \pm i\varepsilon)f) = (\delta_\varepsilon(H_0 - \lambda)f_0, (J^* - T^*R(\lambda \pm i\varepsilon))f).$$

Passing here to the limit $\varepsilon \to 0$ requires of course some assumptions on the perturbation T. We work in the framework of the smooth theory. In conditions of the following assertion, a generalization

$$\Gamma^\pm(\lambda) = \Gamma_0(\lambda)(J^* - T^*R(\lambda \pm i0)), \tag{7.7}$$

of (2.13) is well-defined.

Proposition 7.1 *Under the assumptions of Proposition* 1.18, *for any* $f_0 \in \mathcal{H}_0$, $f \in \mathcal{H}$ *and any compact subinterval* $X \subset \Lambda$,

$$(W^{\pm}(H, H_0; JE_0(X))f_0, f) = \int_X (\Gamma_0(\lambda)f_0, \Gamma^{\pm}(\lambda)f)d\lambda. \tag{7.8}$$

The proof of (7.8) reduces to justification of the passage to the limit $\varepsilon \to 0$ in (7.6) which actually can be done under much weaker assumptions than those given above. Details of this procedure can be found in [165]. Here we mention only that both vector-functions $\Gamma_0(\lambda)f_0$ and $\Gamma^{\pm}(\lambda)f$ in (7.8) are correctly defined for almost all $\lambda \in \Lambda$ and belong to the space $L_2(X; \mathfrak{N})$. For the first of them, this follows from (7.4). To check it for the second, it suffices, by (7.7), to consider (recall that $T = K^*K_0$)

$$(\Gamma_0(\lambda)K_0^*)(KR(\lambda \pm i0)f). \tag{7.9}$$

According to H_0-smoothness of the operator K_0 (see estimate (1.30)),

$$d(K_0E_0(\lambda)K_0^*g, g)/d\lambda \le C||g||^2, \quad \lambda \in X.$$

Hence applying (7.4) to $f = K_0^*g$, we find that the norms $||\Gamma_0(\lambda)K_0^*||$ are uniformly bounded on X. The H-smoothness of the operator K implies that, for any $f \in \mathcal{H}$, the strong limits of $KR(\lambda \pm i\varepsilon)f$ as $\varepsilon \to 0$ exist for almost all $\lambda \in \Lambda$ and the function $||KR(\lambda \pm i0)f||$ belongs to $L_2(X)$. Therefore (7.9) is a product of a bounded and L_2-functions.

Representation (7.8) of the wave operator can again be rewritten in the form (2.23) where the operator $F^{\pm}$ is defined by equality (2.14). The intertwining property of $W^{\pm}(H, H_0; JE_0(X))$ is equivalent to equality (2.17), which implies that, for any $a \in \mathfrak{N}$, elements $\Gamma^{\pm}(\lambda)^*a$ may be considered as eigenfunctions of the continuous spectrum of the operator H, i.e., $H\Gamma^{\pm}(\lambda)^*a = \lambda\Gamma^{\pm}(\lambda)^*a$.

In applications the vector-functions $\Gamma^{\pm}(\lambda)f$ for f from some suitable dense set are usually continuous in λ (away from some exceptional points). For the Schrödinger operator with a short-range potential Proposition 7.1 leads to Theorem 2.9; in this case $J = I$ and we can set $X = \mathbb{R}_+$. A realization of Proposition 7.1 for long-range potentials is given by Theorem 9.5.

7.3. A stationary formula for S-matrix. To describe a stationary formula for $S(\lambda)$ under general circumstances, we need auxiliary wave operators

$$\Omega^{\pm} = W^{\pm}(H_0, H_0; (J^+)^*J^-E_0(\Lambda)) = s - \lim_{t\to\pm\infty} e^{iH_0t}(J^+)^*J^-E_0(\Lambda)e^{-iH_0t}, \tag{7.10}$$

which do not depend on H. The operator $\Omega^{\pm}$ commutes with H_0 and hence $F_0\Omega_{\pm}F_0^*$ acts in the space $L_2(\Lambda; \mathfrak{N})$ as multiplication by the operator-function $\Omega^{\pm}(\lambda) : \mathfrak{N} \to \mathfrak{N}$. The scattering matrix admits two representations which can formally be written as

$$S(\lambda) = \Omega^+(\lambda) - 2\pi i\Gamma_0(\lambda)\Big((J^+)^*T^- - (T^+)^*R(\lambda + i0)T^-\Big)\Gamma_0^*(\lambda) \tag{7.11}$$

and

$$S(\lambda) = \Omega^-(\lambda) - 2\pi i\Gamma_0(\lambda)\Big((T^+)^*J^- - (T^+)^*R(\lambda + i0)T^-\Big)\Gamma_0^*(\lambda) \tag{7.12}$$

with $T^{\pm} = HJ^{\pm} - J^{\pm}H_0$. For definiteness, we give a precise meaning to representation (7.11). The following assertion was obtained in [172].

Proposition 7.2 *Suppose that*

$$T^{\pm} = K^* B^{\pm} K_0 \quad \textit{for both signs and} \quad (J^+)^* T^- = K_0^* \tilde{B} K_0, \tag{7.13}$$

where $B^{\pm}$, $\tilde{B}$ *are bounded operators in some auxiliary Hilbert space* $\mathcal{G}$, $K_0 : \mathcal{H}_0 \to \mathcal{G}$ *is* H_0*-bounded and* $K : \mathcal{H} \to \mathcal{G}$ *is* $|H|^{1/2}$*-bounded. Assume that*

$$Z_0(\lambda; K_0) = \Gamma_0(\lambda) K_0^* : \mathcal{G} \to \mathfrak{N} \tag{7.14}$$

are bounded operators and the operator-function $Z_0^*(\lambda; K_0) : \mathfrak{N} \to \mathcal{G}$ *in strongly continuous in* $\lambda \in \Lambda$. *Let, finally, the operator-function*

$$\mathcal{R}(z; K) = K R(z) K^* : \mathcal{G} \to \mathcal{G} \tag{7.15}$$

be weakly continuous with respect to the parameter z *in the half-band* $\operatorname{Re} z \in \Lambda$, $\operatorname{Im} z \geq 0$. *Then the wave operators* $W^{\pm}(H, H_0; JE_0(\Lambda))$, $W^{\pm}(H_0, H; J^* E(\Lambda))$ *(for* $J = J^+$ *and* $J = J^-$*) and (7.10) exist and the scattering matrix* $S(\lambda)$ *admits for* $\lambda \in \Lambda$ *the representation*

$$S(\lambda) = \Omega^+(\lambda) - 2\pi i Z_0(\lambda; K_0) \tilde{B} Z_0^*(\lambda; K_0) + 2\pi i Z_0(\lambda; K_0)(B^+)^* \mathcal{R}(z; K) B^- Z_0^*(\lambda; K_0). \tag{7.16}$$

In particular, the operator-function $S(\lambda) - \Omega^+(\lambda)$ *is weakly continuous in* $\lambda \in \Lambda$.

Under the assumptions of Proposition 7.2 the operators K_0 and K are H_0- and H-smooth, respectively, on Λ. Therefore both triples H_0, H, J^+ and H_0, H, J^- satisfy the conditions of Proposition 1.18, which implies the existence of all wave operators $W^{\pm}(H, H_0; JE_0(\Lambda))$ and $W^{\pm}(H_0, H; J^* E(\Lambda))$ (both for $J = J^+$ and $J = J^-$). The existence of wave operators (7.10) can be deduced from it with the help of multiplication theorem (1.11). Only equality (7.16) needs to be justified. Formally it is the same as (7.11) but we have a combination of bounded operators in its right-hand side. This gives a correct meaning to (7.11). We usually write representation (7.11) keeping in mind that its precise form is given by (7.16). Its proof is practically the same as the proof of the corresponding assertion in [165] in the case $J^+ = J^-$. Therefore we shall give only a sketch of the proof concentrating on formula representations and omitting details of their justification.

Taking into account the intertwining property of the wave operators

$$W^{\pm}(J) := W^{\pm}(H, H_0; JE_0(\Lambda)),$$

let us rewrite representation (7.8) in the form

$$(E(X) W^{\pm}(J) f_0, f) = \int_X \lim_{\varepsilon \to 0} (\delta_\varepsilon (H_0 - \lambda) f_0, J^* f - T^* R(\lambda \pm i\varepsilon) f) d\lambda, \tag{7.17}$$

where $J = J^+$ or $J = J^-$ and $T = HJ - JH_0$ (so $T = T^+$ or $T = T^-$). By virtue of (1.2), this ensures that

$$\lim_{\varepsilon \to 0} (\delta_\varepsilon (H - \lambda) W^+(J) f_0, f) = \lim_{\varepsilon \to 0} ((J - R(\lambda - i\varepsilon) T) \delta_\varepsilon (H_0 - \lambda) f_0, f) \tag{7.18}$$

for almost all $\lambda \in \Lambda$.

Changing in (7.17) the roles of H_0 and H, we see also that

$$(W^{\pm}(J)f_0, f) = (f_0, W^{\pm}(H_0, H; J^* E(\Lambda))f)$$
$$= (Jf_0, E(\Lambda)f) + \int_\Lambda \lim_{\varepsilon\to 0}(TR_0(\lambda \pm i\varepsilon)f_0, \delta_\varepsilon(H-\lambda)f)d\lambda.$$

In particular, for $J = J^-$ and $f = W^+(J^+)g_0$ we obtain that

$$(W^+(J^-)f_0 - W^-(J^-)f_0, W^+(J^+)g_0)$$
$$= 2\pi i \int_\Lambda \lim_{\varepsilon\to 0}(T^-\delta_\varepsilon(H_0-\lambda)f_0, \delta_\varepsilon(H-\lambda)W^+(J^+)g_0)d\lambda.$$

Equality (7.18) for $J = J^+$ implies (at least formally) that $\delta_\varepsilon(H-\lambda)W^+(J^+)g_0$ in the last integral may be replaced by $(J^+ - R(\lambda - i\varepsilon)T^+)\delta_\varepsilon(H_0-\lambda)g_0$. Hence for any $f_0, g_0 \in \mathcal{H}_0$

$$(W^+(J^-)f_0 - W^-(J^-)f_0, W^+(J^+)g_0)$$
$$= 2\pi i \int_\Lambda \lim_{\varepsilon\to 0}(((J^+)^* - (T^+)^*R(\lambda + i\varepsilon))T^-\delta_\varepsilon(H_0-\lambda)f_0, \delta_\varepsilon(H_0-\lambda)g_0)d\lambda$$
$$= 2\pi i \int_\Lambda (((J^+)^* - (T^+)^*R(\lambda+i0))T^- dE_0(\lambda)f_0/d\lambda, dE_0(\lambda)g_0/d\lambda)d\lambda.$$

The left-hand side here equals

$$(\Omega^+ f_0 - \mathbf{S}(H, H_0; J^+, J^-)f_0, g_0).$$

According to (7.4), the right-hand side may be rewritten as

$$2\pi i \int_\Lambda (\Gamma_0(\lambda)((J^+)^* - (T^+)^*R(\lambda+i0))T^-\Gamma_0^*(\lambda)(F_0f_0)(\lambda), (F_0g_0)(\lambda))d\lambda.$$

This yields representation (7.11).

7.4. Restriction on the energy shell. In the right-hand sides of (7.11) and (7.12), the first terms $\Omega^+(\lambda)$ and $\Omega^-(\lambda)$ do not depend on the operator H. The next terms

$$\Gamma_0(\lambda)(J^+)^*T^-\Gamma_0^*(\lambda) \quad \text{and} \quad \Gamma_0(\lambda)(T^+)^*J^-\Gamma_0^*(\lambda)$$

are also quite explicit and, under the assumptions of Proposition 7.2, they are correctly defined.

Supposing that $\Gamma_0(\lambda)$ is operator (1.21), we consider here the operators of such structure in a more general framework. Namely, given an operator A, we find conditions which guarantee that the restriction $A^\flat(\lambda) = \Gamma_0(\lambda)A\Gamma_0^*(\lambda)$ of the operator A on the energy shell $|\xi| = \lambda^{1/2}$ is well-defined as an operator in the space $\mathfrak{N} = L_2(\mathbb{S}^{d-1})$. If the operator A is regarded as integral operator with kernel $k(\mu,\nu) : \mathfrak{N} \to \mathfrak{N}$ in the direct integral $L_2(\mathbb{R}_+; \mathfrak{N})$ diagonalizing $H_0 = -\Delta$, then $A^\flat(\lambda) = k(\lambda,\lambda)$.

The simplest condition is that $A = \langle x\rangle^{-l}B\langle x\rangle^{-l}$ for some $l > 1/2$ and a bounded operator B. Then, by the Sobolev theorem,

$$A^\flat(\lambda) = (\Gamma_0(\lambda)\langle x\rangle^{-l})B(\Gamma_0(\lambda)\langle x\rangle^{-l})^*$$

is a product of three bounded operators. This remark is sufficient for the short- but not for the long-range case.

In the long-range case the operator A emerges naturally as a PDO. We suppose that its symbol $a(x,\xi)$ belongs to the class $\mathcal{S}^m_{\rho,\delta}$ where $\rho > 1/2 > \delta$. It follows from (1.21) and (2.11) that, formally, $A^\flat(\lambda)$ can be regarded as an integral operator with kernel

$$p^\flat(\omega,\omega';\lambda) = 2^{-1}k^{d-2}(2\pi)^{-d}\int_{\mathbb{R}^d} e^{ik\langle x,\omega'-\omega\rangle}a(x,k\omega')dx, \quad k=\lambda^{1/2}, \tag{7.19}$$

(such integrals are understood, of course, in the sense of distributions). The kernel $p^\flat(\omega,\omega';\lambda)$ is an infinitely differentiable function of ω and ω' (and $\lambda > 0$) for $\omega \neq \omega'$. However due to a possible strong singularity of function (7.19) on the diagonal $\omega = \omega'$ a precise definition of the restriction $A^\flat(\lambda)$ requires some assumptions on the symbol of A. For example, as explained above, it is well-defined as a bounded operator in the space $\mathfrak{N}$ if $a \in \mathcal{S}^m_{\rho,\delta}$ for some $m < -1$.

The operator $A^\flat(\lambda)$ can also be regarded (this point of view was introduced in [20]) as a PDO on $\mathbb{S}^{d-1}$. Its principal symbol $a^\flat(\lambda)$ is invariantly defined on the cotangent bundle of the unit sphere and, according to (7.19), is given by the integral

$$a^\flat(\omega,b;\lambda) = (4\pi k)^{-1}\int_{-\infty}^{\infty} a(\tau\omega + k^{-1}b, k\omega)d\tau, \quad |\omega| = 1, \quad \langle b,\omega\rangle = 0. \tag{7.20}$$

In particular, this integral is absolutely convergent and $a^\flat(\lambda) \in \mathcal{S}^{m+1}_{\rho,\delta}$ if $a \in \mathcal{S}^m_{\rho,\delta}$ for $m < -1$. It was shown in [115] that these assertions remain true for arbitrary m if $a(x,\xi) = 0$ in a conical (in the variable x) neighbourhood of the conormal bundle of the sphere $|\xi| = k$. Note that the principal symbol differs from the total one by a term from the class $\mathcal{S}^{m+2-2\rho}_{\rho,\delta}$; this term depends on chart coordinates. In the critical case $m = -1$, the symbol $a^\flat(\lambda) \in \mathcal{S}^0_{\rho,\delta}$, so the operator $A^\flat$ is bounded. Actually, only this case is needed in Section 9 (as well as in [168] where the scattering matrix of multiparticle systems was studied). Formula (7.20) can be extended (see [174]) to PDO defined by their amplitudes which allows one to dispense with the condition $\rho > 1/2$. This is necessary to treat the general long-range case.

8. THE SHORT-RANGE CASE

A consistent discussion of the scattering matrix and of its spectral properties in the framework of abstract operator theory can be found in [24] and [165]. The survey [24] contains also applications to differential operators with short-range coefficients. We review briefly the short-range case mainly to underscore a drastic difference in properties of the scattering matrix compared to the long-range (Section 9) and multiparticle (Section 14) cases.

8.1. The structure of the scattering matrix. In the most important case $\mathcal{H} = \mathcal{H}_0$, $J^+ = J^- = I$, both representations (7.11) and (7.12) coincide:

$$S(\lambda) = I - 2\pi i \Gamma_0(\lambda)\big(V - VR(\lambda + i0)V\big)\Gamma_0^*(\lambda), \tag{8.1}$$

where of course $V = H - H_0$. Suppose that $V = K^*BK$, where $B = B^*$ is a bounded operator in an auxiliary Hilbert space $\mathcal{G}$, and that the operator-functions $Z_0(\lambda; K)$ and $\mathcal{R}(z; K)$ defined by (7.14) and (7.15) satisfy the assumptions of Proposition 7.2. Then the precise meaning of (8.1) is given by the formula

$$S(\lambda) = I - 2\pi i Z_0(\lambda; K)B(I - \mathcal{R}(\lambda + i0; K)B)Z_0^*(\lambda; K). \tag{8.2}$$

Consider now the two-particle Schrödinger operator $H = -\Delta + V(x)$ in the space $\mathcal{H} = L_2(\mathbb{R}^d)$ with a short-range potential V. By Theorem 2.4, the wave operators $W^{\pm}(H, H_0)$ exist and are complete and hence the scattering matrix $S(\lambda)$ is a unitary operator on the space $L_2(\mathbb{S}^{d-1})$ for almost all $\lambda > 0$. The stationary representation provides an additional information. Recall that, in the case $H_0 = -\Delta$, the operator $\Gamma_0(\lambda)$ was defined by equality (1.21). As shown in subsections 2.1 and 2.2, all conditions of Proposition 7.2 hold for $K_0 = K = \langle x \rangle^{-l}$ where $l = \rho/2 > 1/2$ and $\Lambda = (0, \infty)$. Moreover, both functions $Z_0(\lambda; \langle x \rangle^{-l})$ and $\mathcal{R}(z; \langle x \rangle^{-l})$ are compact and continuous in norm. Thus, we have

Theorem 8.1 *Under the assumptions of Theorem 2.4, the scattering matrix $S(\lambda)$ satisfies for all $\lambda > 0$ representation (8.1) or, more precisely, (8.2). It is a unitary operator on the space $L_2(\mathbb{S}^{d-1})$ and depends continuously (in the topology of the norm) on $\lambda > 0$. The operator $S(\lambda) - I$ is compact and hence its spectrum consists of eigenvalues lying on the unit circle $\mathbb{T}$ and accumulating at the point 1 only.*

Note that representations (8.1) and (1.17), (1.24) for the scattering matrix formally coincide.

As explained in subsection 3.6, it is rather the spectrum of the modified scattering matrix $\Sigma = \Sigma(\lambda)$ defined by (3.18) and (3.19) than of the scattering matrix itself that has a direct interpretation in terms of solutions of the Schrödinger equation. In the spherically symmetric case $V(x) = V(|x|)$ the operators S and Σ have common eigenfunctions (spherical functions) and their eigenvalues may differ by the sign only. However in the general case their eigenfunctions and eigenvalues may be quite different.

The operator $\Sigma - \mathcal{J}$ is compact, ane hence the spectrum of the operator Σ consists of eigenvalues accumulating at the points $\mu = \pm 1$ only.

It follows from the resolvent identity (1.13) and high-energy estimate (2.9) that operator (8.1) admits the following asymptotic expansion (the Born expansion)

$$S(\lambda) = I - 2\pi i \sum_{n=0}^{N} (-1)^n \Gamma_0(\lambda) V (R_0(\lambda + i0)V)^n \Gamma_0^*(\lambda) + G_N(\lambda), \tag{8.3}$$

where

$$||G_N(\lambda)|| = O(\lambda^{-(N+2)/2}) \quad \text{as} \quad \lambda \to \infty.$$

Expansion (8.3) is of course also valid for small perturbations V (and fixed λ). Moreover, under some additional assumptions, (8.3) gives an expansion of the kernel of $S(\lambda) - I$ in a series of increasingly smooth kernels.

According to (2.4), representation (8.2) may be rewritten as

$$S(\lambda) = I - 2\pi i Z_0(\lambda) B (I + \mathcal{R}_0(\lambda + i0)B)^{-1} Z_0^*(\lambda), \tag{8.4}$$

where $Z_0(\lambda) = Z_0(\lambda; K)$ and $\mathcal{R}_0(\lambda + i0) = \mathcal{R}_0(\lambda + i0; K)$. If we choose $K = |V|^{1/2}$, then $B = \operatorname{sgn} V$. Note that the inverse operator in (8.4) exists and

$$(I + \mathcal{R}_0(\lambda + i0)B)^{-1} \in \mathcal{B}. \tag{8.5}$$

In view of (1.2), (7.4), we have that

$$\pi Z_0^*(\lambda) Z_0(\lambda) = \operatorname{Im} \mathcal{R}_0(\lambda + i0). \tag{8.6}$$

Many properties of the scattering matrix hold true for an arbitrary operator of the structure (8.4) provided the inverse operator in its right-hand side exists and identity (8.6) is satisfied. In particular, such operators S are necessarily unitary and $S - I \in \mathfrak{S}_\infty$ if $\operatorname{Im} \mathcal{R}_0 \in \mathfrak{S}_\infty$. Moreover, $S - I \in \mathfrak{S}_p$ if $\operatorname{Im} \mathcal{R}_0 \in \mathfrak{S}_p$.

8.2. Sign-definite perturbations. In some sense eigenvalues of the operators $S(\lambda)$ or $\Sigma(\lambda)$ play on the continuous spectrum the role of eigenvalues of the operator H. From the viewpoint of this analogy, the results of this subsection (see also subsection 10.1 and [164], for details) can be interpreted as a variational principle for the spectrum of $S(\lambda)$ or $\Sigma(\lambda)$.

It is convenient to proceed from general representation (8.4) where we omit dependence of different operators on the parameter λ.

Proposition 8.2 *If $\mathcal{R}_0 \in \mathfrak{S}_\infty$ and $B \geq 0$ ($B \leq 0$), then eigenvalues of the operator* (8.4) *may accumulate at the point $1 \in \mathbb{T}$ only from below (from above).*

Indeed, changing, if necessary, notation we can get $B = \pm I$, so that

$$\operatorname{Im} S = \mp 2\pi Z_0 (I \pm \mathcal{R}_0)^{-1} Z_0^*.$$

Let us represent $\mathcal{R}_0$ as $\mathcal{R}_0 = \mathcal{R}_{00} + \mathcal{K}$ where $||\mathcal{R}_{00}|| < 1$ and $\operatorname{rank} \mathcal{K} < \infty$. Then

$$\Omega = 2\pi Z_0 (I \pm \mathcal{R}_{00})^{-1} Z_0^* \geq 0$$

and the operator $\operatorname{Im} S \pm \Omega$ has finite rank. This implies that eigenvalues of the self-adjoint operator $\operatorname{Im} S$ may accumulate at the point 0 from the left (from the right) only.

Using the spectral theorem for unitary operators, it is easy to deduce from Proposition 8.2 that eigenvalues of the operator $\Sigma = S\mathcal{J}$ may accumulate at the points 1 and -1 in the counterclockwise (clockwise) direction only.

In application to the Schrödinger operator we obtain

Theorem 8.3 *Let estimate* (1.15) *hold for* $\rho > 1$ *and* $V \geq 0$ ($V \leq 0$). *Then eigenvalues of the operators* $S(\lambda)$ *and* $\Sigma(\lambda)$ *may accumulate at the points* 1 *and* ± 1, *respectively, in the counterclockwise* (*clockwise*) *direction only.*

8.3. The asymptotics of scattering phases. Let us write eigenvalues $\mu_n^\pm = \mu_n^\pm(\lambda)$ of $S = S(\lambda)$ as

$$\mu_n^\pm = \exp(\pm 2i\phi_n^\pm), \quad \phi_n^+ \in (0, \pi/2], \ \phi_n^- \in (0, \pi/2), \quad \phi_{n+1}^\pm \leq \phi_n^\pm, \tag{8.7}$$

where the numbers $\phi_n^\pm$ are called the scattering phases. It turns out that the asymptotics of the scattering phases $\phi_n^\pm$ is determined by the asymptotics of the potential $V(x)$ at infinity and is given by the Weyl type formula. Let L_ω be the hyperplane in $\mathbb{R}^d$ orthogonal to $\omega \in \mathbb{S}^{d-1}$. It is assumed that $\mathbb{S}_\omega^{d-2} = \mathbb{S}^{d-1} \cap L_\omega$ is endowed with the usual $(d-2)$-dimensional Euclidean measure if $d > 2$; for $d = 2$ the "sphere" $\mathbb{S}_\omega^{d-2}$ consists of two points, with unit measure assigned to each of them. For a function v on $\mathbb{S}^{d-1}$, set

$$\mathbf{v}(\omega, \psi; \rho) = \int_0^\pi v(\omega \cos\vartheta + \psi \sin\vartheta) \sin^{\rho-2}\vartheta d\vartheta, \quad \psi \in \mathbb{S}_\omega^{d-2}, \tag{8.8}$$

$\mathbf{v}_+ = \max\{\mathbf{v}, 0\}$, $\mathbf{v}_- = \mathbf{v}_+ - \mathbf{v}$ and

$$\mathcal{V}_\pm(v, \rho) = 4^{-1}(d-1)^{-\gamma}(2\pi)^{1-\rho}\Big(\int_{\mathbb{S}^{d-1}} d\omega \int_{\mathbb{S}_\omega^{d-2}} \mathbf{v}_\mp(\omega, \psi; \rho)^{1/\gamma} d\psi\Big)^\gamma \tag{8.9}$$

where $\gamma = (\rho - 1)(d-1)^{-1}$. The following assertion was established in [20].

Theorem 8.4 *Let*

$$V(x) = v(\hat{x})|x|^{-\rho} + o(|x|^{-\rho}), \quad \rho > 1, \ v \in C^\infty(\mathbb{S}^{d-1}), \tag{8.10}$$

as $|x| \to \infty$. *Then the phases* $\phi_n^\pm$ *have the asymptotics*

$$\lim_{n\to\infty} n^\gamma \phi_n^\pm(\lambda) = \lambda^{-1+\rho/2} \mathcal{V}_\pm(v, \rho), \quad \lambda > 0. \tag{8.11}$$

The proof of this theorem [20] relies on the observation that the asymptotics of the phases $\phi_n^{\pm}$ is determined only by the self-adjoint operator $\Gamma_0 V \Gamma_0^*$ (the first Born approximation) in (8.1). Moreover, by calculation of the asymptotics of $\phi_n^{\pm}$ as $n \to \infty$, the potential $V(x)$ can be replaced by its asymptotics $v(\hat{x})|x|^{-\rho}$ as $|x| \to \infty$. According to (7.19), $\Gamma_0 V \Gamma_0^*$ is an integral operator with kernel

$$p^b(\omega, \omega') = 2^{-1} k^{d-2} (2\pi)^{-d} \int_{\mathbb{R}^d} \exp(-ik\langle x, \omega - \omega' \rangle) V(x) dx. \tag{8.12}$$

This kernel has a diagonal singularity of order $|\omega - \omega'|^{\rho - d}$ which determines the asymptotics of eigenvalues of the operator $\Gamma_0 V \Gamma_0^*$. However it is more convenient to regard $\Gamma_0 V \Gamma_0^*$ as a PDO of negative order $1 - \rho$ on the sphere $\mathbb{S}^{d-1}$. By virtue of (7.20), the principal symbol of this operator is

$$a^b(\omega, b) = (4\pi k)^{-1} \int_{-\infty}^{\infty} V(\omega\tau + k^{-1} b) d\tau, \quad b \in L_\omega. \tag{8.13}$$

To compute the leading term in the asymptotics of positive λ_n^+ and negative λ_n^- eigenvalues of PDO $\Gamma_0 V \Gamma_0^*$, one can use the Weyl type formula obtained in [19]. This yields

$$\lim_{n \to \infty} n^{\gamma} \lambda_n^{\pm}(\Gamma_0 V \Gamma_0^*) = \pi^{-1} \lambda^{-1+\rho/2} \mathcal{V}_{\mp}(v, \rho). \tag{8.14}$$

Now it is easy to derive (8.11). Indeed, (8.14) implies the estimate

$$s_n(\Gamma_0 \langle x \rangle^{-l}) = O(n^{-\beta}), \quad \beta = (l - 1/2)(d-1)^{-1}, \quad l > 1/2, \tag{8.15}$$

for singular numbers of the operator $\Gamma_0 \langle x \rangle^{-l}$. Clearly, the principal term in the asymptotics of $2\phi_n^{\pm}$ is the same as that of $\lambda_n^{\pm}(\operatorname{Im} S)$. By virtue of representation (8.1),

$$\operatorname{Im} S = (2i)^{-1}(S - S^*) = -2\pi \Gamma_0 V \Gamma_0^* + \pi \Gamma_0 V (R + R^*) V \Gamma_0^*, \quad R = R(\lambda + i0).$$

It follows from (8.15) and Theorem 2.8 that

$$s_n(\Gamma_0 V R V \Gamma_0^*) = o(n^{-\gamma}). \tag{8.16}$$

Therefore the Weyl theorem on preservation of power asymptotics gives us the asymptotics of $\lambda_n^{\pm}(\operatorname{Im} S)$ and hence (8.11).

If only estimate (1.15) with $\rho > 1$ holds, then

$$\phi_n^{\pm} = O(n^{-\gamma}). \tag{8.17}$$

For the proof, it suffices to combine representation (8.1) with bounds (8.15) and (8.16).

Given an analogy between the scattering phases $\phi_n^{\pm}$ and eigenvalues of the operator H, formula (8.11) plays the role of the Weyl formula for the asymptotics of eigenvalues. In particular, formula (8.11) also has a remarkable generality. Thus, it is preserved (see Section 10) for eigenvalues of the scattering matrix for the pair

(1.29) where V_0 is some short- or long-range potential and V satisfies (8.10). Moreover, it is preserved [25] in the case of a periodic function V_0 if integration in (8.9) over the sphere is replaced by integration over a corresponding energy surface.

8.4. The asymptotics of modified scattering phases. Let us carry over Theorem 8.4 to the modified scattering matrix Σ defined by (3.19). We distribute eigenvalues of Σ over the quadrants and denote by $\hat{\mu}_n^+$ ($\check{\mu}_n^+$) and $\hat{\mu}_n^-$ ($\check{\mu}_n^-$) the eigenvalues of Σ accumulating at the point 1 (−1) clockwise and counterclockwise. Let

$$\hat{\mu}_n^\pm = \exp(\pm 2i\hat{\phi}_n^\pm), \quad \check{\mu}_n^\pm = -\exp(\pm 2i\check{\phi}_n^\pm),$$

with all phases chosen in $(0, \pi/4]$. Modified scattering phases $\hat{\phi}_n^\pm$, $\check{\phi}_n^\pm$, as well as $\phi_n^\pm$ considered in the previous subsection, can be regarded as natural generalizations of the limiting phases arising in scattering by a spherically symmetric potential.

Theorem 8.5 *Let estimate* (1.15) *hold for* $\rho > 1$. *Suppose that the even part*

$$V_e(x) = (V(x) + V(-x))/2$$

of V *satisfies the condition*

$$V_e(x) = v_e(\hat{x})|x|^{-\rho_e} + o(|x|^{-\rho}), \quad \rho > 1, \; v_e \in C^\infty(\mathbf{S}^{d-1}), \quad 2\rho > \rho_e + 1,$$

as $|x| \to \infty$. *Then the following limits exist:*

$$\lim_{n\to\infty} (2n)^{\gamma_e} \hat{\phi}_n^\pm = \lim_{n\to\infty} (2n)^{\gamma_e} \check{\phi}_n^\pm = \lambda^{-1+\rho/2} \mathcal{V}_\pm(v_e, \rho_e), \quad \gamma_e = (\rho_e - 1)(d-1)^{-1}, \quad (8.18)$$

where the function $\mathcal{V}_\pm$ *is defined by* (8.8) *and* (8.9).

The proof of this result is, to a large extent, similar to that of Theorem 8.4. It can be found in [162]. As a by-product of our considerations we obtain also (cf. (8.17)) that

$$\hat{\phi}_n^\pm = O(n^{-\gamma_e}), \quad \check{\phi}_n^\pm = O(n^{-\gamma_e})$$

if $V_e(x) = O(|x|^{-\rho_e})$ and V satisfies (1.15) for $2\rho > \rho_e + 1$.

Thus the asymptotics (and the estimate) of the modified phases is determined only by the even part of V. The odd part of V can decrease slower than V_e (if $\rho_e > \rho$) without contributing to asymptotics (8.18). On the other hand, if $V(x)$ is odd, then Theorem 8.5 gives only that the modified phases are $O(n^{-p})$ where p is any number less than 2γ. The following question remains open.

Problem 8.6 *Let a potential* $V(x) = -V(-x)$ *satisfy the condition* (8.10). *Find the asymptotics of* $\hat{\phi}_n^\pm$ *and* $\check{\phi}_n^\pm$ *as* $n \to \infty$. *This question seems to be open even for a dipole potential*

$$V(x) = v_0(|x + a|^{-1} - |x - a|^{-1}), \quad a \in \mathbf{R}^d.$$

8.5. The scattering cross-section. In scattering experiments one sends a beam of particles of energy λ in direction ω and measures the part $\sigma(\theta,\omega;\lambda)|d\theta|$ of particles scattered in a solid angle $|d\theta|$. This quantity is called the differential scattering cross-section but, abusing somewhat physical terminology, we call the function $\sigma(\theta,\omega;\lambda)$ itself the cross-section in direction θ for the direction ω of an incident beam of particles. According to formula (1.16) (or, more precisely, (3.15)) which is valid for $\rho > (d+1)/2$, the differential cross-section can be expressed via the scattering amplitude by the equality

$$\sigma(\theta,\omega;\lambda) = |a(\theta,\omega;\lambda)|^2, \quad \theta \neq \omega. \tag{8.19}$$

By virtue of (1.24), $a(\theta,\omega;\lambda)$ coincides (up to a numerical factor) for $\theta \neq \omega$ with the kernel $s(\theta,\omega;\lambda)$ of the scattering matrix $S(\lambda)$. This allows us to rewrite (8.19) as

$$\sigma(\theta,\omega;\lambda) = (2\pi)^{d-1}\lambda^{-(d-1)/2}|s(\theta,\omega;\lambda)|^2, \quad \theta \neq \omega. \tag{8.20}$$

Of course $s(\theta,\omega;\lambda)$ can also be understood here as the kernel of $S(\lambda) - I$. Equality (8.20) remains meaningful for all $\rho > 1$ (at least if condition (3.17) is imposed) and, actually, for long-range potentials also. Integrating $\sigma(\theta,\omega;\lambda)$ over all outgoing directions θ, we obtain the total scattering cross-section. By (8.20), this quantity, averaged over incident directions ω, can be expressed in terms of the Hilbert-Schmidt norm $||\cdot||_2$ of the operator $S(\lambda) - I$:

$$\int_{\mathbb{S}^{d-1}}\int_{\mathbb{S}^{d-1}} \sigma(\theta,\omega;\lambda)d\theta d\omega = (2\pi)^{d-1}\lambda^{-(d-1)/2}||S(\lambda) - I||_2^2. \tag{8.21}$$

According to (8.17) this expression is finite if (1.15) is satisfied for $\rho > (d+1)/2$.

Following [146, 147], we shall estimate the difference $S(\lambda) - I$ in arbitrary classes $\mathfrak{S}_p$, $1 \le p < \infty$, where the norm is defined by (1.25). Let us proceed from representation (8.4). If $\mathcal{R}_0 \in \mathfrak{S}_p$, then

$$||S - I||_p \le 2||B||\,||(I + \mathcal{R}_0 B)^{-1}||\,||\operatorname{Im}\mathcal{R}_0||_p, \tag{8.22}$$

so $S - I \in \mathfrak{S}_p$. Estimate (8.22) is useful for small $||\mathcal{R}_0 B||$ only, since otherwise the norm of the inverse operator in (8.22) cannot be controlled. An effective estimate of $||S - I||_p$ can be obtained if unitarity of the operator S is explicitly taken into account.

Theorem 8.7 *Suppose that conditions* (8.5), (8.6) *hold and* $\mathcal{R}_0 \in \mathfrak{S}_p$. *Then operator* (8.4) *satisfies the estimate*

$$||S - I||_p \le c(p)||B||\,||\mathcal{R}_0||_p, \tag{8.23}$$

where

$$c(p) = 2(5 - 2^{1+q})^{1/p}(2^q - 1), \quad q = (p+1)^{-1}.$$

The idea of the proof is to use the estimate

$$s_{m+2k-2}(S - I) \le 2\pi||B||\, s_m((I + \mathcal{R}_0 B)^{-1}) s_k^2(Z_0) \le ||B||(1 - s_m(\mathcal{R}_0 B))^{-1} s_k^2(\operatorname{Im}\mathcal{R}_0)$$

for m large enough; in particular, it is required that $s_m(\mathcal{R}_0 B) < 1$. For first $m-1$ singular numbers we use the trivial estimate $s_n(S-I) \leq 2$. Then it remains to optimize the estimate obtained for $||S-I||_p$ with respect to the choice of m.

We give an application of Theorem 8.7 to the three-dimensional Schrödinger operator only.

Theorem 8.8 *Let $S(\lambda)$ be the scattering matrix for the pair $H_0 = -\Delta$, $H_0 = -\Delta + V$ in the space $L_2(\mathbf{R}^3)$. Then*

$$||S-I||_2 \leq (2\pi)^{-2}(2^{1/2}+1)^4 \int_{\mathbf{R}^3}\int_{\mathbf{R}^3} |V(x)V(y)||x-y|^{-2}dxdy. \tag{8.24}$$

If we introduce a coupling constant g replacing V by gV, then (8.24) shows that the total cross-section is bounded by Cg^2 which is of interest for $g \to \infty$. Estimate (8.24) is in some sense universal. In particular, convergence of the integral in its right-hand side is practically necessary for finiteness of the total cross-section. We emphasize its analogy to the well-known Cwikel-Lieb-Rosenbljum estimate (see, for example, [134], v.4) for eigenvalues of the Schrödinger operator.

Let us consider the opposite case of compactly supported potentials. In classical mechanics (see e.g. [111]) the total cross-section depends only on the size of support. Somewhat similar result is true also in quantum mechanics.

Theorem 8.9 *Under the assumptions of Theorem* 8.8, *suppose additionally that $V(x) \geq 0$ and* $\operatorname{supp} V \subset \{x \in \mathbf{R}^3 : |x| \leq r)$. *Then*

$$||S-I||_2^2 \leq C(\lambda, r).$$

We emphasize that in contrast to classical mechanics the condition $V(x) \geq 0$ cannot be omitted here (see [161], for counterexamples).

The scattering cross-section can also be estimated by time-dependent methods [129]. This approach allows strong local singularities of V but requires averaging over energy intervals.

8.6. The spectral shift function. A study of spectral properties of the scattering matrix $S(\lambda)$ was started by M. S. Birman and M. G. Kreĭn [16, 17] for perturbations of trace-class type. In these papers a relation of $S(\lambda)$ with the Kreĭn spectral shift function $\xi(\lambda)$ was discovered, and the study of these two objects was intimately interrelated. A consistent presentation of the ξ-function theory can be found in the surveys [23], [18] and the book [165]. Here we recall only some basic facts supposing for simplicity that $H - H_0 = V \in \mathfrak{S}_1$.

In the abstract framework the condition of subsection 7.1 that the spectrum of H_0 has constant multiplicity (even on some interval) is unnatural. Therefore relation (7.2) should be replaced by

$$F_0 : \mathcal{H}_0^{(ac)} \to \int_\Lambda^\oplus \mathfrak{N}(\lambda)d\lambda. \tag{8.25}$$

Here Λ (a core of the spectrum of H) is a Borel set of minimal Lebesgue measure such that $E_0(\mathbb{R} \setminus \Lambda) = 0$. The operator F_0 is unitary and satisfies (7.3). The right-hand side of (8.25) is called a direct integral associated with $H_0^{(ac)}$.

The following estimate was obtained in [16]:

$$\int_\Lambda ||S(\lambda) - I_\lambda||_1 d\lambda \le 2\pi ||V||_1. \tag{8.26}$$

In particular,

$$S(\lambda) - I_\lambda \in \mathfrak{S}_1 \tag{8.27}$$

for almost all $\lambda \in \Lambda$. This inclusion follows also from (8.2) since, by (7.4) and Proposition 1.14,

$$Z_0(\lambda; K)^* Z_0(\lambda; K) = dKE_0(\lambda)K^*/d\lambda \in \mathfrak{S}_1$$

for any $K \in \mathfrak{S}_2$. Estimate (8.26) should be compared with (8.23). Both of them give effective bounds for $S - I$. The difference is that the right-hand of (8.26) contains only the trace norm of V whereas the free resolvent R_0 intervenes in (8.23). On the other hand, (8.23) does not require integration over λ and the class $\mathfrak{S}_p$ is arbitrary there.

The spectral shift function $\xi(\lambda) = \xi(\lambda; B, A)$ is introduced by the relation

$$\operatorname{Tr}\big(\varphi(H) - \varphi(H_0)\big) = \int_{-\infty}^{\infty} \varphi'(\lambda)\xi(\lambda) d\lambda, \quad \xi(\lambda) = \xi(\lambda; H, H_0), \tag{8.28}$$

known as the trace formula. The concept of the spectral shift function in the perturbation theory appeared at the beginning of fifties in the physics literature in the papers by I. M. Lifshitz [116]. Its mathematical theory was shortly created by M. G. Kreĭn [103, 104, 105] who proved relation (8.28) for a wide class of functions φ. Moreover, he showed that

$$\int_{-\infty}^{\infty} |\xi(\lambda)| d\lambda \le ||V||_1, \quad \int_{-\infty}^{\infty} \xi(\lambda) d\lambda = \operatorname{Tr} V,$$

$|\xi(\lambda)| \le ||V||$ and $\pm\xi(\lambda) \ge 0$ if $\pm V \ge 0$ for almost all $\lambda \in \mathbb{R}$. In a gap of the continuous spectrum $\xi(\lambda)$ depends on the shift of the eigenvalues of the operator H relative to the eigenvalues of H_0. This explains the term "spectral shift function".

A link between the spectral shift function and the scattering matrix is given by the Birman-Kreĭn formula:

$$\det S(\lambda) = \exp\big(-2\pi i \xi(\lambda)\big) \tag{8.29}$$

for almost all $\lambda \in \Lambda$ (according to (8.27) the determinant of $S(\lambda)$ is correctly defined). This elegant relation is often used for the definition of the spectral shift function on the absolutely continuous part of the spectrum. Formula (8.29) shows that, in the framework of trace class perturbations, results on the spectrum of the scattering matrix are closely related to the theory of the ξ-function.

9. THE LONG-RANGE CASE

Compared to the short-range case, the structure of the scattering matrix $S(\lambda)$ is amazingly different for long-range potentials. In particular, due to the strong diagonal singularity of kernel of $S(\lambda)$, the operator $S(\lambda) - I$ is no longer compact. We give a complete description of this singularity. Its nature is such that the spectrum of $S(\lambda)$ covers the whole unit circle. This section (except subsection 9.7) is based on the paper [172].

9.1. Analytical results. We need some analytical facts on the resolvent $R(z)$ of the Schrödinger operator $H = -\Delta + V$ complementing Theorems 4.2 and 4.3.

Theorem 9.1 *Let* $l > 1/2$ *and* $G_j = \langle x\rangle^{-1/2}\nabla_j^{\perp}$ *with* $\nabla_j^{\perp}$ *defined by* (4.13). *Under the assumptions of Theorem* 4.2,

$$\langle x\rangle^{-l}R(z)G_j^*, \quad G_jR(z)G_k^*, \quad \forall j,k = 1,\dots,d,$$

and continuous (respectively, in the strong and weak topologies) operator-functions of the parameter z *in the region* $\operatorname{Re} z > 0,\ \pm\operatorname{Im} z \geq 0$.

The following resolvent estimates are usually called microlocal or propagation estimates.

Theorem 9.2 *Let condition* (4.1) *on a potential* V *of the operator* $H = -\Delta + V$ *be satisfied for all* κ. *Suppose that the symbol* $a_\pm(x,\xi)$ *of a PDO* $A_\pm$ *belongs to a class* $S_{\rho,\delta}^{-1}$ *for* $\rho > \delta$, $a_\pm(x,\xi) = 0$ *in a neighbourhood of* $\xi = 0$ *and for large* $|\xi|$ *and that the support of* $a_\pm(x,\xi)$ *is contained in the cone*

$$\mp\langle \xi, x\rangle \geq \epsilon|x|\,|\xi|, \quad \epsilon > 0, \quad \hat{\xi} = \xi/|\xi|,\ \hat{x} = x/|x|.$$

Then the operator-functions

$$\langle x\rangle^{l}A_+^*R(z)\langle x\rangle^{-l}, \quad \langle x\rangle^{-l}R(z)A_-\langle x\rangle^{l}, \quad l > 1/2, \tag{9.1}$$

and, for arbitrary p,

$$\langle x\rangle^{p}A_+^*R(z)A_-\langle x\rangle^{p} \tag{9.2}$$

are bounded and continuous in norm with respect to the parameter z *in the region* $\operatorname{Re} z > 0,\ \operatorname{Im} z \geq 0$.

Clearly, symbols of the operators G_j considered as PDO have only a simple zero on the line $x = \gamma\xi$, $\gamma \in \mathbb{R}$. The condition of Theorem 9.2 on supports of symbols $a_\pm$ is much more restrictive but p in (9.2) may be arbitrary. Thus, the results of Theorems 4.3 and 9.1 (the radiation estimates) are intermediary between the limiting absorption principle (Theorem 4.2) and the propagation estimates of Theorem 9.2.

The following result complements the Sobolev trace theorem. Recall that the trace operator $\Gamma_0(\lambda)$ is defined by equality (1.21), its "adjoint" $\Gamma_0^*(\lambda)$ – by equality (2.11) and $\mathfrak{N} = L_2(\mathbf{S}^{d-1})$.

Proposition 9.3 *The operator-functions $G_j\Gamma_0^*(\lambda) : \mathfrak{N} \to \mathcal{H}$ are bounded and (as well as their adjoints $\Gamma_0(\lambda)G_j^*$) are strongly continuous in $\lambda > 0$.*

Proofs of Theorem 9.1 and Proposition 9.3 are given in [168]. Proofs of Theorem 9.2 can be found either in [121, 90, 88] or in [81, 84]. The proof of [121, 90, 88] relies on the Mourre estimate and is easily accessible.

We need also a result on the essential spectrum of PDO with oscillating symbols. In view of our applications we consider PDO on the unit sphere $\mathbf{S}^{d-1}$ but the problem reduces by a diffeomorphism to operators acting in a domain $\Omega \subset \mathbf{R}^{d-1}$.

Proposition 9.4 *Let A be a PDO with symbol from the class $S^0_{\rho,\delta}$, where $\rho > 1/2 > \delta$. Suppose that for some point (ω_0, b_0), $|\omega_0| = 1$, $\langle\omega_0, b_0\rangle = 0$, $b_0 \neq 0$, the principal symbol $a(\omega, b)$ of A admits the representation*

$$a(\omega_0, tb_0) = e^{i\theta(t)}(1 + o(1)), \quad t \to \infty,$$

where $\theta(t)$ is a continuous function and $|\theta(t_n)| \to \infty$ for some sequence $t_n \to \infty$. Then the spectrum of the operator A in the space $\mathfrak{N}$ covers the unit circle.

The proof of Proposition 9.4 can be found in [172]. Its generalization to arbitrary $\rho > 0$ is given in [174].

9.2. Stationary representations. As in Section 4, we consider for simplicity only potentials V satisfying (4.1) for $\rho > 1/2$ and all κ. Let $J^\pm$ be PDO (4.21) with symbol (4.26), (4.28). The existence, isometricity and completeness of the operators $W^\pm = W^\pm(H, H_0; J^\pm)$ were verified in Theorem 4.7.

A stationary representation of $W^\pm(H, H_0; J^\pm)$ can be deduced from Proposition 7.1. Let K be a "vector" operator with components $\langle x\rangle^{-l}$, $l > 1/2$, and $G_j = \langle x\rangle^{-1/2}\nabla_j^\perp$. Recall that "perturbation" (4.22) satisfies representation (4.32), which gives factorization $T^\pm = K^*B^\pm K$ with a bounded operator $B^\pm$. By Theorems 4.2 and 4.3, the operator K is H_0- and H-smooth. Moreover, by Theorem 9.1 and Proposition 9.3, the operator-functions $\langle x\rangle^{-l}R(z)K^*$ and $K\Gamma_0^*(\lambda)$ are strongly continuous in z and λ, respectively. Thus we obtain

Theorem 9.5 *Let condition (4.1) hold for $\rho > 1/2$ and all κ. Then the operator*

$$\Gamma^\pm(\lambda) = \Gamma_0(\lambda)((J^\pm)^* - (T^\pm)^*R(\lambda \pm i0)) : L_2^{(l)}(\mathbf{R}^d) \to \mathfrak{N}, \quad l > 1/2, \tag{9.3}$$

is bounded for any $l > 1/2$ and depends continuously in the weak sense (its adjoint is strongly continuous) on $\lambda > 0$. The operator $F^\pm$ defined by formula (2.14) on the set $L_2^{(l)}(\mathbf{R}^d)$ extends by continuity to a bounded operator $F^\pm : L_2(\mathbf{R}^d) \to L_2(\mathbf{R}_+; \mathfrak{N})$, and it satisfies equalities (2.16), (2.17) and (2.19). The wave operators $W^\pm$ are related to $F^\pm$ by formulas (2.22) or (2.23).

Our study of the scattering matrix (7.1) relies on Proposition 7.2. Remark that, again by Theorem 9.1, the operator-function $KR(z)K^*$ is weakly continuous in z. Therefore it remains to check factorization (7.13) for the operator $(J^+)^*T^-$. By virtue of (4.31) the symbol t_r^- of the regular part T_r^- of the perturbation belongs to the class $S^{-2\rho}_{\rho,1-\rho}$ and, consequently, the same is true for the symbol of the PDO $(J^+)^*T_r^-$. The symbol t_s^- of the singular part T_s^- is defined by equalities (4.29) and (4.30). Comparing it with (4.28), we see that the principal symbol of the PDO $(J^+)^*T_s^-$ is

$$g_0(x,\xi) = -2ie^{i\Phi(x,\xi)}\eta^2(x)\psi^2(|\xi|^2)\sigma^+(\langle\hat{x},\hat{\xi}\rangle)\langle\xi,\nabla\sigma^-(\langle\hat{x},\hat{\xi}\rangle)\rangle, \tag{9.4}$$

where

$$\Phi(x,\xi) = \Phi^-(x,\xi) - \Phi^+(x,\xi) = 2^{-1}\int_{-\infty}^{\infty}\Big(V(s\xi) - V(x+s\xi)\Big)ds. \tag{9.5}$$

This leads to

Lemma 9.6 *The operator $(J^+)^*T^-$ is a PDO with symbol $\mathbf{g}(x,\xi)$ which differs from $g_0(x,\xi)$ by a function from the class $S^{-2\rho}_{\rho,1-\rho}$. The function $g_0 \in S^{-1}_{\rho,1-\rho}$ and $g_0(x,\xi) = 0$ in a neighbourhood of the cotangent bundle $\hat{\xi} = \pm\hat{x}$ to the sphere $|\xi| = \mathrm{const}$.*

Hence representation (4.32) for the operator $(J^+)^*T^-$ can be obtained exactly in the same way as that for $T^\pm$ in subsection 4.3.

Note also that according to (4.36)

$$s - \lim_{t\to\pm\infty} J^\mp \exp(-iH_0t) = 0,$$

so $\Omega^\pm = 0$ and hence $\Omega^\pm(\lambda) = 0$ for all $\lambda > 0$.

Thus the scattering matrix (7.1) admits representation (7.11), where $\Omega^+(\lambda) = 0$. We can assert that every term

$$S_1(\lambda) = -2\pi i\Gamma_0(\lambda)(J^+)^*T^-\Gamma_0^*(\lambda), \tag{9.6}$$

$$S_2(\lambda) = 2\pi i\Gamma_0(\lambda)(T^+)^*R(\lambda+i0)T^-\Gamma_0^*(\lambda) \tag{9.7}$$

in the right-hand side of (7.11) is bounded and is a weakly continuous function of $\lambda > 0$. Moreover, their sum $S(\lambda)$, being unitary, is strongly continuous in $\lambda > 0$. Let us summarize the results obtained.

Theorem 9.7 *In conditions of Theorem 9.5, the scattering matrix (7.1) equals*

$$S(\lambda) = S_1(\lambda) + S_2(\lambda) \tag{9.8}$$

with S_1, S_2 defined by (9.6), (9.7). The function $S(\lambda)$ is strongly continuous in $\lambda > 0$.

Theorems 9.5 and 9.7 can be extended to the general case discussed in subsection 4.5.

We emphasize that, in contrast to short-range potentials, the scattering matrix is not continuous in norm in the long-range case (even for the Coulomb potential $V(x) = v|x|^{-1}$); see the end of subsection 9.7.

A representation of $S(\lambda)$ to a large extent similar to that of Theorem 9.7 appeared first in [83]. However it seems to us that (9.6) was not well defined there as a bounded operator in $\mathfrak{N}$. Indeed, its definition requires either Theorem 9.1 or results of the type [115] mentioned in subsection 7.4, but such assertions were not used in [83].

9.3. The structure of the scattering matrix. We emphasize that operators (9.6) and (9.7) depend on the choice of the cut-off functions $\sigma^\pm$ and η in definition (4.21), (4.28) of the identifications $J^\pm$, but their sum (9.8) does not depend on it. Below we suppose that, for some $\epsilon \in (0,1)$,

$$\sigma^+(\vartheta) = 1 \text{ if } \vartheta \in [-\epsilon, 1] \quad \text{and} \quad \sigma^-(\vartheta) = 1 \text{ if } \vartheta \in [-1, \epsilon]. \tag{9.9}$$

Then the term $S_2(\lambda)$ is in some sense negligible.

Lemma 9.8 *The operator $S_2(\lambda)$ is compact and norm-continuous in λ for $\lambda > 0$.*

To check this result, recall that $T^\pm = T_r^\pm + T_s^\pm$, where $T_r^\pm$, $T_s^\pm$ are PDO with symbols $t_r^\pm$, $t_s^\pm$ defined by (4.29). Let us write (9.7) as

$$S_2(\lambda) = 2\pi i \big(\Gamma_0(\lambda)\langle x\rangle^{-l}\big)\mathfrak{R}(\lambda + i0)\big(\langle x\rangle^{-l}\Gamma_0^*(\lambda)\big), \tag{9.10}$$

where $l \in (1/2, \rho]$,

$$\mathfrak{R}(z) = \mathfrak{R}_{rr}(z) + \mathfrak{R}_{rs}(z) + \mathfrak{R}_{sr}(z) + \mathfrak{R}_{ss}(z),$$

$$\mathfrak{R}_{rr}(z) = \langle x\rangle^l (T_r^+)^* R(z) T_r^- \langle x\rangle^l,$$

$$\mathfrak{R}_{rs}(z) = \langle x\rangle^l (T_r^+)^* R(z) T_s^- \langle x\rangle^l, \quad \mathfrak{R}_{sr}(z) = \langle x\rangle^l (T_s^+)^* R(z) T_r^- \langle x\rangle^l,$$

$$\mathfrak{R}_{ss}(z) = \langle x\rangle^l (T_s^+)^* R(z) T_s^- \langle x\rangle^l.$$

In view of estimates (4.31), the operators $\langle x\rangle^l T_r^\pm \langle x\rangle^l$ are bounded. Therefore, applying Theorem 4.2, we find that the operator $\mathfrak{R}_{rr}(z)$ is bounded and norm-continuous in z for $\operatorname{Re} z > 0$, $\operatorname{Im} z \geq 0$. According to (4.30) and (9.9), the support of the symbol $t_s^+(x,\xi)$ belongs to the cone $\langle \hat{x}, \hat{\xi}\rangle \leq -\epsilon$ and, similarly, the support of the symbol $t_s^-(x,\xi)$ belongs to the cone $\langle \hat{x}, \hat{\xi}\rangle \geq \epsilon$. Since, moreover, $t_s^\pm \in \mathcal{S}_{\rho,1-\rho}^{-1}$, Theorem 9.2 can be applied to the operators $\mathfrak{R}_{rs}(z)$, $\mathfrak{R}_{sr}(z)$ (see (9.1)) and $\mathfrak{R}_{ss}(z)$ (see (9.2)). It follows that the operator-function $\mathfrak{R}(z)$ is bounded and norm-continuous in z for $\operatorname{Re} z > 0$, $\operatorname{Im} z \geq 0$. To conclude the proof of Lemma 9.8 we return to representation (9.10) and use that the operator $\Gamma_0(\lambda)\langle x\rangle^{-l} : \mathcal{H} \to \mathfrak{N}$ is compact and norm-continuous in λ.

Let us now consider operator (9.6). The following result follows from Lemma 9.6.

Lemma 9.9 *Let G_0 be the PDO with symbol* (9.4) *and*

$$S_0(\lambda) = -2\pi i \Gamma_0(\lambda) G_0 \Gamma_0^*(\lambda). \tag{9.11}$$

Then the operator $S(\lambda) - S_0(\lambda)$ is compact and norm-continuous in λ for $\lambda > 0$.

It remains to study operator (9.11). Under assumption (9.9) the function $\sigma^+(\langle\hat{\xi}, \hat{x}\rangle)$ equals 1 on the support of $(\sigma^-)'(\langle\hat{\xi}, \hat{x}\rangle)$ and hence it may be omitted in (9.4). Since $(\sigma^-)'(\langle\hat{x}, \omega\rangle) = 0$ in neighbourhoods of the points $\hat{x} = \pm\omega$, the principal symbol $\mathbf{s}(\omega, b; \lambda)$ of PDO (9.11) can be calculated by formula (7.20), which yields

$$\mathbf{s}(\omega, b; \lambda) = -i(2k)^{-1} \int_{-\infty}^{\infty} g_0(\tau\omega + k^{-1}b, k\omega) d\tau, \quad |\omega| = 1, \quad \langle b, \omega\rangle = 0, \quad k = \lambda^{1/2}. \tag{9.12}$$

Note that the operator $S_0(\lambda)$ is defined by its principal symbol (9.12) up to a term from the class $\mathcal{S}_{\rho,\delta}^{-2\rho+1}$, which is compact. To calculate integral (9.12) remark that, by definition (9.5), the function

$$\Phi(\tau\omega + k^{-1}b, k\omega) = \Phi(k^{-1}b, k\omega) = k^{-1}\Phi(k^{-1}b, \omega)$$

does not, actually, depend on τ. Note also that for $\xi = k\omega$, $x = \tau\omega + k^{-1}b$

$$\langle \xi, \nabla\sigma^-(\langle\hat{\xi}, \hat{x}\rangle)\rangle = k\partial\sigma^-(\tau(\tau^2 + k^{-2}b^2)^{-1/2})/\partial\tau.$$

The integral of this function over τ equals $-k$. According to (9.12) this implies that

$$\mathbf{s}(\omega, b; \lambda) = \exp\bigl(ik^{-1}\Phi(k^{-1}b, \omega)\bigr), \quad |\omega| = 1, \quad \langle b, \omega\rangle = 0. \tag{9.13}$$

Naturally, the principal symbol of PDO (9.11) does not depend on the cut-off functions $\sigma^\pm$ although expression (9.4) for the principal symbol of G_0 depends on them.

Thus we have proven

Theorem 9.10 *Under assumptions* (4.1) *for $\rho > 1/2$ and all κ and* (9.9),

$$S(\lambda) = S_0(\lambda) + \tilde{S}(\lambda),$$

where $S_0(\lambda)$ is a PDO on $\mathbb{S}^{d-1}$ with the principal symbol (9.5), (9.13) *and the operator $\tilde{S}(\lambda)$ is compact.*

The operator $S_0(\lambda)$ is strongly continuous in $\lambda > 0$. Indeed, the symbol of the derivative $S_0'(\lambda)$ belongs to the class $\mathcal{S}_{\rho,\delta}^{1-\rho}$ and hence $||S_0'(\lambda)a||$ is locally bounded in λ on a suitable dense set of elements a. Therefore $S_0(\lambda)a$ is a strongly continuous function of λ, which together with the uniform boundedness of $||S_0(\lambda)||$ implies the strong continuity of $S_0(\lambda)$. Since the function $\tilde{S}(\lambda)$ is norm-continuous, we thus recover the result of Theorem 9.7 on strong continuity of the scattering matrix.

Combining Theorem 9.10 with Proposition 9.4, we obtain

Theorem 9.11 *Let the assumptions of Theorem* 9.5 *hold. Suppose that function* (9.5) *satisfies the condition*

$$|\Phi(t_n b_0, \omega_0)| \to \infty \tag{9.14}$$

for some point $\omega_0 \in \mathbb{S}^{d-1}$, $\langle b_0, \omega_0\rangle = 0$, $b_0 \neq 0$, and some sequence $t_n \to \infty$. Then for all $\lambda > 0$ the spectrum of the scattering matrix $S(\lambda)$ coincides with the unit circle.

Remark that condition (9.14) is satisfied for all points $\omega \in \mathbb{S}^{d-1}$, $\langle b,\omega\rangle = 0$, $b \neq 0$, if $V(x)$ is an asymptotically homogeneous function of order $-\rho$, $\rho \in (1/2,1]$. Theorems 9.7, 9.10 and 9.11 can be extended to V satisfying (4.1) for arbitrary $\rho > 0$ and to potentials which are sums of a long-range and of a short-range terms. On the other hand, in the short-range case $\rho > 1$ we have that

$$\lim_{|b|\to\infty} \Phi(b,\omega) = 0, \quad \forall\omega \in \mathbb{S}^{d-1}, \quad \langle b,\omega\rangle = 0.$$

Therefore the principal symbol of S equals 1 which corresponds to the Dirac-function in its kernel. In the long-range case this singularity disappears.

As was advocated in subsection 3.5, in the short-range case eigenvalues of the modified scattering matrix Σ defined by equalities (3.18), (3.19) have a more direct physical interpretation than eigenvalues of S. This leads to

Problem 9.12 *Describe the spectrum of the operator Σ in the long-range case.*

We do not have any information on the structure of the spectrum of the scattering matrix. Note, however, that in the radial case $V(x) = V(|x|)$ it consists of eigenvalues. By Theorem 9.11, they are dense on the unit circle; see subsection 9.7 for more details.

9.4. The scattering amplitude. Our study of the integral kernel $s(\omega,\omega';\lambda)$ of $S(\lambda)$ (of the scattering amplitude) relies again on the representation (9.6) – (9.8) where the operator $J^\pm$ is PDO (4.21). However compared to (4.28) its symbol

$$\mathrm{j}^\pm(x,\xi) = e^{i\Phi^\pm(x,\xi)} a^\pm(x,\xi)\eta(x)\sigma^\pm(\langle\hat\xi,\hat x\rangle)\psi(|\xi|^2) \tag{9.15}$$

will contain an (approximate) solution $a^\pm(x,\xi)$ of the transport equation

$$2\langle\xi,\nabla a^\pm\rangle = -(i|\nabla\Phi^\pm|^2 + \Delta\Phi^\pm)a^\pm - 2\langle\nabla\Phi^\pm,\nabla a^\pm\rangle + i\Delta a^\pm.$$

A meaning of this construction is that, compared to $e^{i\varphi^\pm}$, the functions $e^{i\varphi^\pm}a^\pm$ are a better approximation to eigenfunctions of H. We recall that the functions $\varphi^\pm$ and $\Phi^\pm$ are connected by equality (4.23).

By iterations, we obtain the following expression

$$a^\pm(x,\xi) = a_N^\pm(x,\xi) = \sum_{n=0}^{N} b_n^\pm(x,\xi), \tag{9.16}$$

where $b_0^\pm = 1$ and $b_n^\pm$ are determined inductively by relations:

$$q_0^\pm = i|\nabla\Phi^\pm|^2 + \Delta\Phi^\pm, \quad q_n^\pm = 2\langle\nabla\Phi^\pm,\nabla b_n^\pm\rangle + q_0^\pm b_n^\pm - i\Delta b_n^\pm, \quad n \geq 1, \tag{9.17}$$

$$b_{n+1}^\pm(x,\xi) = \pm 2^{-1}\int_0^\infty q_n^\pm(x \pm \tau\xi,\xi)d\tau, \tag{9.18}$$

so that $2\langle\xi,\nabla b_{n+1}^\pm\rangle + q_n^\pm = 0$ (cf. (4.26)). Then instead of (4.24) we get the equality

$$(-\Delta + V - |\xi|^2)(e^{i\varphi^\pm}a^\pm) = -ie^{i\varphi^\pm}q^\pm, \quad q^\pm = q_N^\pm.$$

The following elementary assertion shows that, for a proper choice of N, the error $q_N^\pm$ decays at infinity faster than any power of $|x|^{-1}$.

Lemma 9.13 *For any $\nu \in (-1,1)$, in the region $\pm\langle \hat{\xi}, \hat{x}\rangle \geq \pm\nu$*

$$|\partial_x^\alpha \partial_\xi^\beta b_n^\pm(x,\xi)| \leq C_{\alpha,\beta}(1+|x|)^{-\varepsilon_0 n-|\alpha|}, \quad \varepsilon_0 = 2\rho - 1 > 0, \tag{9.19}$$

and

$$|\partial_x^\alpha \partial_\xi^\beta q^\pm(x,\xi)| \leq C_{\alpha,\beta}(1+|x|)^{-1-\varepsilon_0(N+1)-|\alpha|}. \tag{9.20}$$

Let $J_N^\pm$ be the PDO with symbol (9.15) corresponding to $a = a_N^\pm$. It follows from (9.16) and (9.19) that the difference $J_N^\pm - J_0^\pm$ is compact for any N. Therefore, for all N,

$$W^\pm(H, H_0; J_N^\pm) = W^\pm(H, H_0; J_0^\pm) \tag{9.21}$$

where $W^\pm(H, H_0; J_0^\pm)$ is the wave operator constructed in Section 4. Note also operator (9.3) is not changed if the operators $J^\pm = J_0^\pm$ and $T^\pm = T_0^\pm$ in the right-hand side are replaced by $J_N^\pm$ and $T_N^\pm = HJ_N^\pm - J_N^\pm H_0$, respectively.

Equality (9.21) implies that the scattering matrices $S(\lambda; H, H_0; J_N^+, J_N^-)$ also coincide for different N. If we replace $J^\pm = J_0^\pm$ and $T^\pm = T_0^\pm$ in (9.6) and (9.7) by $J_N^\pm$ and $T_N^\pm$, then each term $S_1(\lambda) = S_1^{(N)}(\lambda)$ and $S_2(\lambda) = S_2^{(N)}(\lambda)$, separately, might depend on N. However, representation (9.8) is preserved.

Quite straightforwardly (cf. subsection 4.3), we find that the perturbation $T_N^\pm$ is PDO with symbol (4.29), where however

$$\tau_s^\pm = -2ia^\pm\langle\nabla\varphi^\pm, \nabla\sigma^\pm\rangle\eta - 2\langle\nabla a^\pm, \nabla\sigma^\pm\rangle\eta - a^\pm\Delta\sigma^\pm\eta, \quad \sigma^\pm = \sigma^\pm(\langle\hat{x},\hat{\xi}\rangle), \tag{9.22}$$

(instead of (4.30)) and $\tau_r^\pm = \eta q^\pm\sigma^\pm$, up to terms containing $\nabla\eta$ or $\Delta\eta$, so that

$$|\partial_x^\alpha \partial_\xi^\beta \tau_r^\pm(x,\xi)| \leq C_{\alpha,\beta}(1+|x|)^{-1-\varepsilon_0(N+1)-|\alpha|} \tag{9.23}$$

(instead of (4.31)). As before, $T_r^\pm$, $T_s^\pm$ are PDO with symbols $t_r^\pm$, $t_s^\pm$ defined by (4.29).

Using estimate (9.23), it is easy to show that the regular part $S_2^{(N)}(\lambda)$ of the scattering matrix $S(\lambda)$ is an integral operator with smooth kernel. The proof of the following assertion is similar to that of Lemma 9.8.

Lemma 9.14 *The operator $S_2^{(N)}$ is an integral operator with kernel*

$$s_2^{(N)} \in C^M(\mathbb{S}^{d-1}\times\mathbb{S}^{d-1}), \quad \text{where} \quad M < (2\rho-1)(N+1) - d + 1.$$

Actually, $s_2^{(N)}(\lambda)$ is also M times differentiable in the energy λ but we fix λ and omit it from notation.

Let us proceed to analysis of the singular part $S_1^{(N)}(\lambda)$ of the scattering matrix. Combining Lemma 9.13 with standard calculus of PDO we obtain (cf. Lemma 9.6)

Lemma 9.15 *The operator $(J_N^+)^* T_N^-$ is the PDO with symbol*

$$\mathbf{g}_N(x,\xi) = e^{i\Phi(x,\xi)}\mathbf{w}_N(x,\xi) + u_N(x,\xi), \tag{9.24}$$

where $\Phi(x,\xi)$ *is given by* (9.5),

$$\mathbf{w}_N(x,\xi) = -2i\eta^2(x)\psi^2(|\xi|^2)\langle\xi, \nabla\sigma^-(\langle\hat{x},\hat{\xi}\rangle)\rangle + w_N(x,\xi) \tag{9.25}$$

and

$$w_N \in \mathcal{S}_{1,0}^{-2\rho}, \quad u_N \in \mathcal{S}_{\rho,1-\rho}^{-1-(N+1)\varepsilon_0}, \quad \varepsilon_0 = 2\rho - 1. \tag{9.26}$$

According to equality (9.6) the kernel $s_1^{(N)}(\omega,\omega')$ of the operator $S_1^{(N)}$ is determined by formula (7.19). Moreover, the second inclusion (9.26) implies that the term u_N in the right-hand side of (9.24) is negligible. Taking also into account Lemma 9.14, we get

Proposition 9.16 *Let assumption* (4.1) *with* $\rho > 1/2$ *hold. Set, for* $\omega \neq \omega'$,

$$s_0(\omega,\omega';\lambda) = -i2^{-1}(2\pi)^{-d+1}k^{d-2}\int_{\mathbb{R}^d} e^{ik\langle x,\omega'-\omega\rangle}e^{i\Phi(x,k\omega')}\mathbf{w}(x,k\omega')dx, \tag{9.27}$$

where $k = \lambda^{1/2}$, Φ *and* $\mathbf{w} = \mathbf{w}_N$ *are defined by* (9.5) *and* (9.25), *respectively. Then*

$$s - s_0 \in C^M(\mathbb{S}^{d-1} \times \mathbb{S}^{d-1}),$$

where M *is the same number as in Lemma* 9.14.

Remark that the function $e^{i\Phi}\mathbf{w}$ belongs to the class $\mathcal{S}_{\rho,1-\rho}^{-1}$. Thus, if $\omega \neq \omega'$, one can integrate by parts in (9.27) arbitrary number of times. Hence, $s_0(\omega,\omega')$ is an infinitely differentiable function of ω,ω' for $\omega \neq \omega'$. Using now Proposition 9.16 and taking into account that M is arbitrary there, we arrive at the following assertion obtained in [2, 83].

Proposition 9.17 *Under assumption* (4.1), *where* $\rho > 1/2$, *the kernel* $s(\omega,\omega')$ *of the scattering matrix* S *is an infinitely differentiable function of* ω *and* ω' *for* $\omega \neq \omega'$.

9.5. The diagonal singularity. Representation (9.27) allows us to find the diagonal singularity of the kernel $s(\omega,\omega')$. Below we fix the vector $\omega' =: \omega_0$ and study $s_0(\omega,\omega_0)$ as $\omega \to \omega_0$. We set $\tau = \langle\omega_0, x\rangle$, $b = x - \tau\omega_0$ (so that b belongs to the hyperplane L_{ω_0} orthogonal to ω_0) and introduce the vector

$$\vartheta = \omega - \langle\omega,\omega_0\rangle\omega_0 \in L_{\omega_0}. \tag{9.28}$$

By virtue of relations

$$\langle\omega_0 - \omega, b\rangle = -\langle\vartheta, b\rangle, \quad \langle\omega_0 - \omega, \omega_0\rangle = 1 - (1 - |\vartheta|^2)^{1/2} =: f(\vartheta) = O(|\vartheta|^2)$$

for $\vartheta \to 0$, oscillation of the integrand in (9.27) in the variable τ can be neglected. Thus we rewrite (9.27) as

$$s_0(\omega,\omega_0;\lambda) = (2\pi)^{-d+1}k^{d-1}\int_{\mathbb{R}^{d-1}} e^{-ik\langle b,\vartheta\rangle}e^{ik^{-1}\Phi(b,\omega_0)}p(b,\vartheta)db, \tag{9.29}$$

where Φ is given by (9.5) and

$$p(b,\vartheta) = -i2^{-1}k^{-1}\int_{-\infty}^{\infty} \mathbf{w}(\tau\omega_0 + b, k\omega_0)e^{ikf(\vartheta)\tau}d\tau.$$

It can easily be checked that

$$p(b,\vartheta) = P_0(f(\vartheta)b) + p_1(b,\vartheta), \tag{9.30}$$

where

$$P_0(u) = -\int_{-\infty}^{\infty} e^{iku\tau}\partial\sigma^-(\tau(\tau^2+1)^{-1/2})/\partial\tau\, d\tau \tag{9.31}$$

(so $P_0 \in \mathcal{S}(\mathbb{R})$ and $P(0) = 1$) and, for all α,

$$|\partial_b^\alpha p_1(b,\vartheta)| \le C_\alpha(1+|b|)^{-\varepsilon_0-|\alpha|}, \quad \varepsilon_0 = 2\rho - 1 \in (0,1), \tag{9.32}$$

with constants C_α not depending on ϑ.

Integral (9.29) contains all singularities of kernel $s(\omega,\omega';\lambda)$ of the scattering matrix. This result is perhaps of interest even in the short-range case $\rho > 1$. For short-range potentials we can set $\Phi = 0$ which simplifies expression (9.29). Formula (9.29) is of course much more efficient than expansion (8.3). A similar asymptotic formula for $s(\omega,\omega';\lambda)$ was derived by M. M. Skriganov [145] by a different method. It was also shown there that such an asymptotics remains valid for $\lambda \to \infty$. Still earlier the scattering amplitude $s(\omega,\omega';\lambda)$ was studied by V. S. Buslaev [26] for potentials from the class $\mathcal{S}(\mathbb{R}^d)$. In this case the scattering amplitude is a smooth function of ω and ω'. In [26] a complete asymptotic expansion of $s(\omega,\omega';\lambda)$ as $\lambda \to \infty$ was obtained.

In the long-range case $\Phi(b,\omega) \to \infty$ as $|b| \to \infty$, $\langle\omega,b\rangle = 0$, so the asymptotics of integral (9.29) as $\vartheta \to 0$ can be obtained by the stationary phase method. To describe explicitly the leading singularity of $s(\omega,\omega_0)$ as $\omega \to \omega_0$, we assume for simplicity that $V(x)$ is a homogeneous function for sufficiently large $|x|$:

$$V(x) = V_0(x) := v(\hat{x})|x|^{-\rho}, \quad v \in C^\infty(\mathbb{S}^{d-1}),\ |x| \ge r_0. \tag{9.33}$$

Then, in the case $\rho < 1$, function $\Phi(b,\omega)$ is homogeneous, up to some constant $\nu(\omega)$, of degree $1-\rho$ for sufficiently large $|b|$:

$$\Phi(b,\omega) = \mathbf{v}(\hat{b},\omega)|b|^{1-\rho} + \nu(\omega), \quad |b| \ge r_0, \tag{9.34}$$

where

$$\mathbf{v}(\hat{b},\omega) = 2^{-1}\int_{-\infty}^{\infty}\Big(V_0(\tau\omega) - V_0(\hat{b}+\tau\omega)\Big)d\tau \tag{9.35}$$

and

$$\nu(\omega) = 2^{-1}\int_{-r}^{r}\Big(V(\tau\omega) - V_0(\tau\omega)\Big)d\tau, \quad r \ge r_0,$$

does not depend on b (and r).

Making the change of variables $b = (k^2|\vartheta|)^{-\gamma}y$, $\gamma = \rho^{-1}$, we see that the asymptotics of integral (9.29) is determined by stationary points of the function

$$\mathbf{u}(y,\hat{\vartheta},\omega_0) = -\langle y,\hat{\vartheta}\rangle + \mathbf{v}(\hat{y},\omega_0)|y|^{1-\rho}. \tag{9.36}$$

According to (9.35), the equation $\nabla_y \mathbf{u}(y,\hat{\vartheta},\omega_0) = 0$ is equivalent to

$$2\hat{\vartheta} + \int_{-\infty}^{\infty}(\nabla V_0)(y+\tau\omega_0)d\tau = 0, \quad y \in L_{\omega_0}. \tag{9.37}$$

Let us denote by $\mathcal{H}(y,\omega_0)$ the Hessian of the function $\Phi(y,\omega_0)$, $y \in L_{\omega_0}$, i.e., $\mathcal{H}(y,\omega_0)$ is the $(d-1)\times(d-1)$ - matrix with elements

$$\mathcal{H}_{ij}(y,\omega_0) = -2^{-1}\int_{-\infty}^{\infty} \partial^2 V_0(y+\tau\omega_0)/\partial y_i \partial y_j \, d\tau, \quad y \in L_{\omega_0}.$$

Set also

$$\mathfrak{h}(y,\omega_0) = |\det \mathcal{H}(y,\omega_0)|^{-1/2} \exp\big(i\pi \operatorname{sgn} \mathcal{H}(y,\omega_0)/4\big). \tag{9.38}$$

By (9.36), the Hessian of $\mathbf{u}(y,\hat{\vartheta},\omega_0)$ equals $\mathcal{H}(\omega_0,y)$. Moreover, by virtue of (9.30) - (9.32), $p(k^2|\vartheta|^{-\gamma}y,\vartheta) \to 1$ for $\gamma < 2$ and fixed y as $\vartheta \to 0$. This allows us to replace the function $p(b,\vartheta)$ in (9.29) by 1. Omitting simple calculations we formulate the final result.

Theorem 9.18 *Let assumptions* (4.1) *and* (9.33) *with* $\rho \in (1/2,1)$ *hold. Fix* $k > 0$, $\omega_0 \in \mathbb{S}^{d-1}$ *and let* ω *and* ϑ *be related by* (9.28). *Suppose that for a given* $\hat{\vartheta}$ *there is a finite number of points* $y_1(\hat{\vartheta}),\dots,y_n(\hat{\vartheta})$ *satisfying* (9.37) *and that* $\det \mathcal{H}(y_j(\hat{\vartheta}),\omega_0) \neq 0$ *for all* $j=1,\dots,n$. *Define the functions* $\mathbf{u}$ *and* $\mathfrak{h}$ *by equalities* (9.36) *and* (9.38), *respectively. Then the kernel of the scattering matrix admits as* $\omega \to \omega_0$ *or, equivalently,* $\vartheta \to 0$ *the representation*

$$s(\omega,\omega_0;\lambda) = (2\pi k^{2\gamma-1})^{-(d-1)/2}|\vartheta|^{-(d-1)(1+\gamma)/2}\exp(ik^{-1}\nu(\omega_0))$$
$$\times \sum_{j=1}^{n} \mathfrak{h}(y_j(\hat{\vartheta}),\omega_0)\exp\big(ik^{1-2\gamma}|\vartheta|^{1-\gamma}\mathbf{u}(y_j(\hat{\vartheta}),\hat{\vartheta},\omega_0)\big)(1+O(|\vartheta|^{\varepsilon})), \tag{9.39}$$

where $\gamma = \rho^{-1}$, $\varepsilon = \varepsilon(\rho) > 0$.

We emphasize that the moduli and the phases of different terms in (9.39) are asymptotically homogeneous functions, as $\omega - \omega_0 \to 0$, of orders $-(d-1)(1+\rho^{-1})/2 < -d+1$ and $1-\rho^{-1}$, respectively. Thus S is more singular than the singular integral operator. Nevertheless, oscillations of $s(\omega,\omega_0)$ compensate the growth of $|s(\omega,\omega_0)|$, and S is a bounded operator in $L_2(\mathbb{S}^{d-1})$. It is possible that for some $\hat{\vartheta}$ there are no points y satisfying (9.37). In this case $s_0(\omega,\omega_0) \to 0$ as $\vartheta \to 0$ faster than any power of $|\vartheta|$, and hence the kernel $s(\omega,\omega_0)$ of the scattering matrix remains bounded.

Let us consider an example of asymptotically central potentials where $v(\hat{x}) = v = const$ in (9.33). Then the leading term of $s(\omega,\omega_0)$ as $\omega \to \omega_0$ is a function of $|\vartheta|$ only. In this case the sum in (9.39) consists of one term $\mathfrak{h}\exp(ik^{1-2\gamma}|\vartheta|^{1-\gamma}\mathbf{u})$, where

$$\mathfrak{h} = e^{i\pi(d-3)\operatorname{sgn} v/4}2^{-(d-1)/2}\rho^{-1/2}(q(\rho)|v|)^{\gamma(d-1)/2}, \quad q(\rho) = \pi^{1/2}\Gamma((1+\rho)/2)\Gamma(\rho/2)^{-1},$$

(Γ is the gamma-function) and

$$\mathbf{u} = \rho(1-\rho)^{-1}(q(\rho)|v|)^{\gamma}\operatorname{sgn} v.$$

We emphasize that the differential cross-section (8.20) grows as $|\vartheta|^{-(d-1)(1+\gamma)}$ as $\vartheta \to 0$. An example of an essentially non-central potential where $v(\hat{x}) = \langle \hat{x}, n\rangle$ for some given vector $n \in \mathbb{R}^d$ can be found in [172].

9.6. The Coulomb case. The case $\rho = 1$ is essentially different. Now instead of (9.34) function (9.5) admits for $|\omega| = 1$, $\langle b, \omega\rangle = 0$ the representation

$$\Phi(b,\omega) = \mathbf{v}(\omega)\ln|b| + \mu(\hat{b},\omega) + \mu_0(\omega), \quad |b| \geq r_0, \tag{9.40}$$

where $\mathbf{v}(\omega) = (v(\omega) + v(-\omega))/2$,

$$\mu(\omega,\hat{b}) = 2^{-1}\int_{|\tau|\geq 1}\Big(V_0(\tau\omega) - V_0(\hat{b}+\tau\omega)\Big)d\tau - 2^{-1}\int_{|\tau|\leq 1} V_0(\hat{b}+\tau\omega)d\tau,$$

$$\mu_0(\omega) = 2^{-1}\int_{-r}^{r} V(\tau\omega)d\tau - \mathbf{v}(\omega)\ln r, \quad r \geq r_0.$$

To calculate the asymptotics as $|\vartheta| \to 0$ of the integral (9.29), we change the variable $b = |\vartheta|^{-1}y$, use (9.40) and show again that, in the limit, the term $p(|\vartheta|^{-1}y, \vartheta)$ can be replaced by 1. The final result is formulated as follows.

Theorem 9.19 *Let assumptions* (4.1) *and* (9.33) *with* $\rho = 1$ *hold. Fix* $k > 0$, $\omega_0 \in \mathbb{S}^{d-1}$ *and let* ω *and* ϑ *be related by* (9.28). *Set*

$$G(y,\hat{\vartheta},\omega_0,k) = -k\langle y,\hat{\vartheta}\rangle + k^{-1}\mathbf{v}(\omega_0)\ln|y| + k^{-1}\mu(\hat{y},\omega_0).$$

and

$$g(\hat{\vartheta},\omega_0,k) = \exp\Big(ik^{-1}\mu_0(\omega_0)\Big)(2\pi)^{-d+1}k^{d-1}\int_{\mathbb{R}^{d-1}}\exp\Big(iG(y,\hat{\vartheta},\omega_0,k)\Big)dy$$

(*this integral is understood of course in the sense of distributions*). *Then the kernel of the scattering matrix admits as* $\omega \to \omega_0$ *or, equivalently,* $\vartheta \to 0$ *the representation*

$$s(\omega,\omega_0) = g(\hat{\vartheta},\omega_0,k)|\vartheta|^{-d+1-i\mathbf{v}(\omega_0)/k}(1 + O(|\vartheta|^{\epsilon})), \quad \forall\epsilon < 1. \tag{9.41}$$

According to (9.41), the modulus of $s(\omega,\omega_0)$ is asymptotically homogeneous of order $-d+1$, and the phase has a logarithmic singularity on the diagonal. In contrast to a usual singular integral operator, the operator S is bounded in $L_2(\mathbb{S}^{d-1})$ due to oscillations of its kernel $s(\omega,\omega_0)$.

For the Coulomb potential $V(x) = v|x|^{-1}$ the scattering matrix $S(\lambda)$ can be calculated explicitly (see, e.g., [112], [63], [171]). Formally, $S(\lambda)$ is an integral operator with kernel

$$s(\omega,\omega';\lambda) = 2^{i\alpha}\pi^{-\delta}\Gamma(\delta+i\alpha)\Gamma^{-1}(-i\alpha)|\omega-\omega'|^{-d+1-2i\alpha}, \quad \alpha = v(2k)^{-1},\ \delta = (d-1)/2. \tag{9.42}$$

To be more precise, $S(\lambda)$ is the strong limit of operators $S_\varepsilon(\lambda)$ whose kernels are defined by formula (9.42) with $-i\alpha$ replaced by $-i\alpha+\varepsilon$, $\varepsilon > 0$. Let θ be the angle between the vectors ω and ω'. It follows from (9.42) that the differential cross-section (8.20) is given by the Gordon and Mott formula

$$\sigma(\theta;\lambda) = (2k)^{-d+1}|\Gamma(\delta+i\alpha)\Gamma^{-1}(-i\alpha)|^2\sin^{-2d+2}(\theta/2). \tag{9.43}$$

In the classical mechanics (see, e.g., [111]) the differential cross-section satisfies the Rutherford formula

$$\Sigma_{cl}(\theta;\lambda) = (4\lambda)^{-d+1}|v|^{d-1}\sin^{-2d+2}(\theta/2). \tag{9.44}$$

The right-hand sides of (9.43) and (9.44) contain the same function $\sin^{-2d+2}(\theta/2)$ of the scattering angle θ but the numerical coefficients coincide if and only if $d=3$. Note, however, that $|\alpha|^{-d+1}|\Gamma(\delta+i\alpha)\Gamma^{-1}(-i\alpha)|^2 \to 1$ as $|\alpha|\to\infty$ for any d. Thus, the quantum and classical cross-sections of scattering by the Coulomb potential coincide in the quasi-classical limit for all dimensions d.

9.7. Radial potentials. For radial potentials $V(x) = V(|x|)$, the spectrum of the scattering matrix $S = S(\lambda)$ is pure point. Its eigenfunctions are spherical functions. Thus properties of S are determined by its eigenvalues $\mu_m = e^{2i\delta_m}$ which we parametrize by the orbital quantum number $m = 0,1,2,\ldots$.

As in the short-range case (see subsection 3.6), the phases δ_m can be constructed in terms of the asymptotics as $r\to\infty$ of the regular solution $\psi_m(r) = \psi_m(r,\lambda)$ of the radial Schrödinger equation (3.22) fixed by its asymptotics $\psi_m(r) \sim r^{m+(d-1)/2}$ as $r\to 0$. However, in contrast to (3.21) asymptotics of ψ_m depends on the potential V. Actually, it can be easily proved that, for each m, (3.22) has a solution with the asymptotics

$$f_m(r,\lambda) = \exp\Big(i\int_{r_0}^{r}(\lambda - V(s))^{1/2}ds\Big) + o(1) \tag{9.45}$$

as $r\to\infty$. Here r_0 is an arbitrary fixed point common for all m, and we may suppose that, for λ considered, $V(r)\le\lambda$ for $r\ge r_0$. The function $\overline{f_m(r,\lambda)}$ is another solution of (3.22) and the Wronskian

$$\{f_m,\bar f_m\} = f_m'\bar f_m - f_m\bar f_m{}' = 2i\lambda^{1/2}.$$

Clearly,

$$\psi_m(r) = (2i\lambda^{1/2})^{-1}\Big(\{\psi_m,\bar f_m\}f_m(r) - \{\psi_m, f_m\}\bar f_m(r)\Big). \tag{9.46}$$

Set

$$b_m = |\{\psi_m, f_m\}|,\quad \{\psi_m,f_m\} = b_m e^{-i\delta_m} i^m e^{\pi i(d-3)/4}. \tag{9.47}$$

It follows from (9.45), (9.46) that

$$\psi_m(r,\lambda) = \lambda^{-1/2}b_m(\lambda)\sin\Big(i\int_{r_0}^{r}(\lambda - V(s))^{1/2}ds - \pi(2m+d-3)/4 + \delta_m(\lambda)\Big) + o(1)$$

as $r\to\infty$. Of course, the phases $\delta_m(\lambda)$ depend on the choice of r_0 which corresponds to non-uniqueness of the scattering matrix for long-range potentials. Moreover, δ_m are determined up to a multiple of 2π; we fix this number by our convenience.

Let us expound briefly a method for calculating the asymptotics of $\delta_m(\lambda)$ as $m\to\infty$. For large m, solutions of equations (3.22) are described by quasi-classical formulas. This means that, under suitable assumptions on V, equation (3.22) has a solution $y_m(r,\lambda)$ admitting the representation

$$y_m(r,\lambda) = Q(r,m,\lambda)^{-1/4}\exp\Big(i\int_{r_0}^{r}Q(s,m,\lambda)^{1/2}ds\Big)u_m(r,\lambda), \tag{9.48}$$

where

$$Q(r,m,\lambda) = \lambda - p_m r^{-2} - V(r), \quad p_m = (m + (d-2)/2)^2 - 1/4,$$

$\arg Q = 0$ if $Q > 0$ and $\arg Q = \pi$ if $Q < 0$. The function $u_m(r,\lambda) \to 1$ and $u'_m(r,\lambda) \to 0$ as $r + m \to \infty$. Furthermore, one can easily verify that $\psi_m(r) = r^{m+(d-1)/2}\Psi_m(r)$ where $\Psi_m(r)$ and $\Psi'_m(r)$ are bounded as $m \to \infty$. Using (9.48) and calculating the Wronskian at the point $r = r_0$, we see that

$$\{\psi_m, y_m\} = 2e^{-\pi i/4} r_0^{m+(d-2)/2} \Psi_m(r_0) m^{1/2}(1 + o(1)) \tag{9.49}$$

(we do not dwell on estimating remainders).

Comparing (9.45) and (9.48) for $r \to \infty$, we find a relation between the solutions $f_m(r,\lambda)$ and $y_m(r,\lambda)$:

$$f_m(r,\lambda) = \lambda^{1/4} e^{-iI_m(\lambda)} y_m(r,\lambda),$$

where

$$I_m(\lambda) = \int_{r_0}^{\infty} \left((\lambda - p_m r^{-2} - V(r))^{1/2} - (\lambda - V(r))^{1/2} \right) dr. \tag{9.50}$$

Hence it follows from (9.49) that

$$\{\psi_m, f_m\} = 2e^{-\pi i/4} r_0^{m+(d-2)/2} \Psi_m(r_0) \lambda^{1/4} e^{-iI_m(\lambda)} m^{1/2}(1 + o(1)),$$

where $\Psi_m(r_0) > 0$ for sufficiently large m. Thus according to definition (9.47)

$$\delta_m(\lambda) = \operatorname{Re} I_m(\lambda) + \pi(2m + d - 2)/4 + o(1).$$

Assuming that $V(r) = vr^{-\rho}$ for large r, we can easily calculate the asymptotics of integral (9.50) as $m \to \infty$. Let us only formulate the final result:

$$\delta_m(\lambda) = c(\lambda) m^{1-\rho}(1 + o(1)), \quad \rho < 1, \tag{9.51}$$

where

$$c(\lambda) = -2^{-1} v \lambda^{(\rho-2)/2} \int_1^{\infty} x^{-\rho}((1 - x^{-2})^{-1/2} - 1) dx. \tag{9.52}$$

The last integral can be calculated in terms of the Γ-function. If $V(r) = vr^{-1}$ (the Coulomb potential), then $\delta_m(\lambda)$ grows logarithmically. Actually, in this case $\delta_m(\lambda)$ can be calculated (see, e.g., [112]) explicitly:

$$\delta_m(\lambda) = \arg \Gamma(m + 1 + i2^{-1}\lambda^{-1/2} v). \tag{9.53}$$

It follows from (9.51) or (9.53) that eigenvalues of S are dense on the unit circle. This is of course consistent with Theorem 9.11.

Formula (9.51) remains valid for short-range potentials when $\rho > 1$ and hence $\delta_m(\lambda) \to 0$ as $m \to \infty$. In this case the asymptotic coefficient is determined by equality (9.52) where, however, the integrand should be replaced by $x^{-\rho}(1 - x^{-2})^{-1/2}$. Of course this result coincides with formulas (8.9), (8.11) for radial potentials (multiplicities of eigenvalues should be taken into account).

Formulas (9.51) and (9.53) imply that for long-range potentials the function $S(\lambda)$ cannot be continuous in the sense of the norm. Indeed,

$$||S(\lambda) - S(\lambda')|| = \sup_m |e^{2i\delta_m(\lambda)} - e^{2i\delta_m(\lambda')}|$$

and, for any $|\lambda - \lambda'|$, the right-hand side is not small because of large m.

10. THE RELATIVE SCATTERING MATRIX

Here we consider a perturbation of an operator $H_0 = -\Delta + V_0$ by a short-range potential V satisfying assumption (1.15) with some $\rho > 1$. The background potential V_0 may be both short- or long-range. Our aim here is to study the scattering matrix S for the pair H_0, $H = H_0 + V$. The main result of this section is that the principal symbol of $S - I$ (considered as a PDO) does not depend on V_0, and hence it is the same as that for $V_0 = 0$. This result is formulated in Theorem 10.9 and is new.

In this section we set $H_{00} = -\Delta$, and all objects related to H_{00} are supplied with the index "00"; for example, operator (1.21) is denoted now $\Gamma_{00}(\lambda)$.

10.1. Generalities. Combining the limiting absorption principle (see Theorem 4.2) valid for both operators H_0 and H with Proposition 1.18 we obtain

Theorem 10.1 *Suppose that V_0 satisfies the assumptions of Theorem* 4.2 *and V satisfies* (1.15) *where $\rho > 1$. Then the wave operators $W^{\pm}(H, H_0)$ exist and are complete, and the scattering operator $\mathbf{S}(H, H_0)$ is unitary on the subspace $\mathcal{H}_0^{(ac)} = E_0(0,\infty)\mathcal{H}$.*

Given an unitary operator F_0 diagonalizing $H_0^{(ac)} = H_0 E_0(0,\infty)$, the scattering matrix $S(\lambda; H, H_0)$ is defined by formula (7.5). The operator $S(\lambda; H, H_0)$ satisfies stationary representation (8.1) where $\Gamma_0(\lambda)$ is "operator" (2.12). In view of (1.2) and (7.4)

$$||\Gamma_0(\lambda)\langle x\rangle^{-l} f||^2 = \pi^{-1} \operatorname{Im}(\langle x\rangle^{-l} R_0(\lambda + i0)\langle x\rangle^{-l} f, f), \quad l > 1/2, \tag{10.1}$$

and hence $\langle x\rangle^{-l}\Gamma_0^*(\lambda) : L_2(\mathbb{S}^{d-1}) \to \mathcal{H}$ is well-defined as a bounded operator. According to Proposition 7.2 for justification of (8.1) one needs only to check that this operator-function is strongly continuous in $\lambda > 0$. Two possible choices of $\Gamma_0(\lambda) = \Gamma_0^{\pm}(\lambda)$ were suggested in Section 5; see definition (5.6). By Proposition 5.4, the condition of strong continuity is satisfied in these cases.

Theorem 9.5 (see also the paragraph after Lemma 9.13) applied to the operator $H_0 = H_{00} + V_0$ gives another representation

$$\Gamma_0^{\pm}(\lambda) = \Gamma_{00}(\lambda)((J_0^{\pm})^* - (T_0^{\pm})^* R_0(\lambda \pm i0)), \quad T_0^{\pm} = H_0 J_0^{\pm} - J_0^{\pm} H_{00}, \tag{10.2}$$

for the same operator $\Gamma_0^{\pm}(\lambda)$. Here $J_0^{\pm}$ is the PDO with symbol

$$\mathrm{j}_0^{\pm}(x,\xi) = e^{i\Phi_0^{\pm}(x,\xi)} a_0^{\pm}(x,\xi)\eta(x)\sigma^{\pm}(\langle\hat{\xi},\hat{x}\rangle)\psi(|\xi|^2), \tag{10.3}$$

where

$$\Phi_0^{\pm}(x,\xi) = \pm 2^{-1}\int_0^\infty \big(V_0(x \pm s\xi) - V_0(\pm s\xi)\big)ds \tag{10.4}$$

and $a_0^\pm(x,\xi)$ is defined by equalities (9.16) – (9.18) with Φ replaced by Φ_0. The number N in (9.16) may be arbitrary; in particular, $a_0 = 1$ if $N = 0$. As before, it is assumed that $\psi \in C_0^\infty(\mathbb{R}_+)$ and $\psi(\lambda) = 1$ on some compact interval $\Lambda \subset (0,\infty)$. Then (10.2) is satisfied for $\lambda \in \Lambda$. We emphasize that the operator $\Gamma_0^\pm(\lambda)$ does not depend on the choice of the cut-off functions $\sigma^\pm$, η, ψ and the number N. Theorem 9.5 requires that

$$|D^\kappa V_0(x)| \leq C_\kappa (1+|x|)^{-\rho_0-|\kappa|} \tag{10.5}$$

for $\rho_0 > 1/2$ and all κ, but (10.2) can be extended to the general case discussed in subsection 4.5. If V_0 is a short-range function, then $\Gamma_0^\pm(\lambda)$ may be defined by formula (2.13); in this case the operator-function $\langle x\rangle^{-l}\Gamma_0^\pm(\lambda)^*$, $l > 1/2$, is continuous in norm.

The operator-function $\langle x\rangle^{-l}R_0(z)\langle x\rangle^{-l}$ is compact for $\operatorname{Im} z \neq 0$, and by Theorem 4.2, for $l > 1/2$, it is norm-continuous up to the cut along $[0,\infty)$ (except possibly the point $z = 0$). By virtue of (10.1), this implies that the operators $\Gamma_0(\lambda)\langle x\rangle^{-l}$ are compact. This gives us a generalization of Theorem 8.1.

Theorem 10.2 *Let V satisfy* (1.15) *where $\rho > 1$.*

1^0 *If V_0 is also a short-range function, then all conclusions of Theorem* 8.1 *hold for the scattering matrix $S(\lambda) = S(\lambda; H, H_0)$.*

2^0 *If V_0 satisfies the assumptions of Theorem* 4.2, *then conclusions of Theorem* 8.1 *hold with continuity of $S(\lambda)$ in the sense of the norm replaced by the strong one.*

Some results of Section 8 follow immediately from representation (8.1). For example, we have a direct generalization of Theorem 8.3.

Theorem 10.3 *Let V_0 and V satisfy the assumptions of Theorem* 10.1 *and $V \geq 0$ ($V \leq 0$). Then eigenvalues of the operators $S(\lambda)$ and $\Sigma(\lambda) = S(\lambda)\mathcal{J}$ may accumulate at the points* 1 *and* ± 1, *respectively, in the counterclockwise (clockwise) direction only.*

The following result is close in spirit to Theorem 10.3 (see [101] and [164]).

Theorem 10.4 *Let V_0 and V satisfy the assumptions of Theorem* 10.3. *Set $H_\varepsilon = H_0 + \varepsilon V$. Then eigenvalues of the operators $S_\varepsilon(\lambda) = S(\lambda; H_\varepsilon, H_0)$ and $\Sigma_\varepsilon(\lambda) = S_\varepsilon(\lambda)\mathcal{J}$ rotate in the clockwise (counterclockwise) direction as ε increases.*

10.2. The multiplication theorem. Our main goal is to find the principal symbol of the operator of $S(\lambda; H, H_0) - I$ considered as PDO. In contrast to the special case $V_0 = 0$, representation (8.1) is not convenient for this purpose because expressions of Sections 5 and 9 for $\Gamma_0(\lambda)$ (or even (2.13) in the case of short-range V_0) are not efficient enough.

Therefore we are obliged to proceed from an expression for $S(\lambda; H, H_0)$ in terms of scattering matrices for the pairs H_{00}, H_0 and H_{00}, H. We accept now assumption (10.5) and denote by $J^\pm$ the PDO with symbol (10.3). By Theorem 4.7, the wave operators $W^\pm(H_0, H_{00}; J_0^\pm)$ and $W^\pm(H, H_{00}; J_0^\pm)$ exist and, considered as mappings of $E_{00}(\Lambda)\mathcal{H}$ on $E_0(\Lambda)\mathcal{H}$ and $E(\Lambda)\mathcal{H}$, respectively, they are unitary.

The chain rule (1.11) implies that

$$W^{\pm}(H, H_{00}; J_0^{\pm}) = W^{\pm}(H, H_0) W^{\pm}(H_0, H_{00}; J_0^{\pm})$$

and hence, by definition (1.12),

$$\mathbf{S}(H, H_{00}; J_0^+, J_0^-) = W^+(H_0, H_{00}; J_0^+)^* \mathbf{S}(H, H_0) W^-(H_0, H_{00}; J_0^-). \tag{10.6}$$

It is convenient to introduce the operators

$$\mathbf{S}^{\pm}(H, H_0) = W^{\pm}(H_0, H_{00}; J_0^{\pm})^* \mathbf{S}(H, H_0) W^{\pm}(H_0, H_{00}; J_0^{\pm}) \tag{10.7}$$

commuting with H_{00}. The operators $\mathbf{S}^{\pm}(H, H_0) E_{00}(\Lambda)$ and $\mathbf{S}(H, H_0) E_0(\Lambda)$ are unitarily equivalent. Comparing (10.6) and (10.7) we obtain two equalities

$$\mathbf{S}^+(H, H_0) = \mathbf{S}(H, H_{00}; J_0^+, J_0^-) \mathbf{S}(H_0, H_{00}; J_0^+, J_0^-)^* \tag{10.8}$$

and

$$\mathbf{S}^-(H, H_0) = \mathbf{S}(H_0, H_{00}; J_0^+, J_0^-)^* \mathbf{S}(H, H_{00}; J_0^+, J_0^-), \tag{10.9}$$

where all operators are restricted on the subspace $E_{00}(\Lambda)\mathcal{H}$.

As usual, we suppose that H_{00} is diagonalized by formulas (1.21), (2.12) (where Γ_0 and F_0 should be denoted Γ_{00} and F_{00}). The operator $F_{00}\mathbf{S}^{\pm}(H, H_0)F_{00}^*$ acts in the space $L_2(\mathbb{R}_+, L_2(\mathbb{S}^{d-1}))$ as multiplication by operator-function $S^{\pm}(\lambda) = S^{\pm}(\lambda; H, H_0) : L_2(\mathbb{S}^{d-1}) \to L_2(\mathbb{S}^{d-1})$. Both operators $S^{\pm}(\lambda)$ can be regarded as scattering matrices for the pair H_0, H referred to different diagonalizations of the operator H_0. Of course, $S^+(\lambda)$ and $S^-(\lambda)$ for $\lambda \in \Lambda$ are unitary equivalent. In terms of scattering matrices equations (10.8) and (10.9) can be reformulated as follows.

Proposition 10.5 *Let assumptions* (10.5) *for some* $\rho_0 > 1/2$ *and all* κ *and* (1.15) *for some* $\rho > 1$ *be satisfied. Denote by* $S_0(\lambda)$ *and* $S_1(\lambda)$ *the scattering matrices corresponding to the scattering operators*

$$\mathbf{S}(H_0, H_{00}; J_0^+, J_0^-) \quad \text{and} \quad \mathbf{S}(H, H_{00}; J_0^+, J_0^-),$$

respectively. Then the scattering matrices corresponding to operators (10.7) *satisfy the equations*

$$S^+(\lambda) = S_1(\lambda) S_0^*(\lambda), \quad S^-(\lambda) = S_0^*(\lambda) S_1(\lambda), \quad \lambda \in \Lambda. \tag{10.10}$$

We emphasize that equalities (10.8) and (10.9) as well as Proposition 10.5 are actually of abstract nature. Only the existence and isometricity of different wave operators are really required for their validity. In particular, Proposition 10.5 extends to the general case discussed in subsection 4.5. Equalities (10.8), (10.9) and (10.10) can be regarded as a multiplication theorem for scattering operators and scattering matrices, respectively. Note that if V_0 is a short-range function, then these equalities are valid for $J_0^{\pm} = I$ and all $\lambda > 0$.

Considered up to a unitary equivalence, the scattering matrix $S(\lambda; H, H_0)$ can be identified with any of the operators $S^+(\lambda)$ or $S^-(\lambda)$. Both equalities (10.10) can be used for its study. Namely, applying representations (9.6) – (9.8) to $S_0(\lambda)$ and $S_1(\lambda)$, we immediately obtain a representation for $S^{\pm}(\lambda)$.

10.3. The principal symbol. Now we are able to find the principal symbol of the operator $S^{\pm}$. In order to use the PDO calculus, we need however to assume that V satisfies condition (4.1) for $\rho > 1$ and all κ.

Let $J_0^{\pm}$ be PDO with symbol (10.3). We proceed from Theorem 9.7 and suppose that condition (9.9) is satisfied. Then, according to Lemma 9.14, for sufficiently large N in definition (9.16) of function $a_0^{\pm}$, the kernel of the operator

$$\Gamma_{00}(\lambda)(T_0^+)^* R_0(\lambda + i0) T_0^- \Gamma_{00}^*(\lambda), \quad T_0^{\pm} = H_0 J_0^{\pm} - J_0^{\pm} H_{00},$$

is arbitrarily smooth. Therefore the leading singularity of the scattering matrix $S_0(\lambda)$ is determined by the term

$$-2\pi i \Gamma_{00}(\lambda)(J_0^+)^* T_0^- \Gamma_{00}^*(\lambda). \tag{10.11}$$

To obtain a similar result for the scattering matrix $S_1(\lambda)$, we have to introduce a new identification $J^{\pm}$. We choose $J^{\pm}$ as the PDO with symbol $\mathbf{j}^{\pm}$ defined by formula (9.15) where $\Phi^{\pm} = \Phi_0^{\pm}$ and $a^{\pm}$ is constructed by formulas (9.16), (9.18) and

$$q_0^{\pm} = i|\nabla\Phi^{\pm}|^2 + \Delta\Phi^{\pm} + iV, \quad q_n^{\pm} = 2\langle \nabla\Phi^{\pm}, \nabla b_n^{\pm}\rangle + q_0^{\pm} b_n^{\pm} - i\Delta b_n^{\pm}, \quad n \geq 1,$$

(instead of (9.17)). It is easy to check that the remainder

$$q^{\pm} = ie^{-i\varphi^{\pm}}(-\Delta + V_0 + V - |\xi|^2)(e^{i\varphi^{\pm}} a^{\pm}), \quad \varphi^{\pm}(x,\xi) = \langle x, \xi\rangle + \Phi^{\pm}(x,\xi),$$

satisfies estimate (9.20). Under condition (9.9), this allows us to verify, quite similarly to Lemma 9.14, that for sufficiently large N the kernel of the operator

$$\Gamma_{00}(\lambda)(T^+)^* R(\lambda + i0) T^- \Gamma_{00}^*(\lambda), \quad T^{\pm} = HJ^{\pm} - J^{\pm} H_{00},$$

is arbitrarily smooth. Applying now Theorem 9.7 to the pair H_{00}, H with identifications $J^{\pm}$, we see that the leading singularity of the scattering matrix $S(\lambda; H, H_{00}; J^+, J^-)$ is given by the term

$$-2\pi i \Gamma_{00}(\lambda)(J^+)^* T^- \Gamma_{00}^*(\lambda). \tag{10.12}$$

Choosing the same parameter N in the definitions of the functions $a_0^{\pm}$ and $a^{\pm}$, we see that they differ only by terms containing V. Therefore

$$a^{\pm}(x,\xi) - a_0^{\pm}(x,\xi) = \pm i 2^{-1} \int_0^{\infty} V(x \pm s\xi)\, ds \tag{10.13}$$

with an error from the class $\mathcal{S}_{1,0}^{-\rho+1-\varepsilon}$, $\varepsilon > 0$ (in any cone $\pm\langle \hat{\xi}, \hat{x}\rangle \geq \pm\nu$, $\nu \in (-1,1)$). It particular, this implies that the operators $J^{\pm} - J_0^{\pm}$ are compact. Hence

$$\mathbf{S}(H, H_{00}; J^+, J^-) = \mathbf{S}(H, H_{00}; J_0^+, J_0^-)$$

and

$$S_1(\lambda) = S(\lambda; H, H_{00}; J^+, J^-).$$

Let us formulate an intermediary result.

Lemma 10.6 *For any M, we can choose $N = N(M, \rho_0, \rho)$ such that the scattering matrices $S_0(\lambda)$ and $S_1(\lambda)$ coincide with operators* (10.11) *and* (10.12), *respectively, up to operators with kernels from the class $C^M(\mathbb{S}^{d-1} \times \mathbb{S}^{d-1})$.*

By Lemma 10.6, up to negligible terms,

$$-(2\pi i)^{-1}(S_1(\lambda) - S_0(\lambda)) = \Gamma_{00}(\lambda)((J^+)^* T^- - (J_0^+)^* T_0^-)\Gamma_{00}^*(\lambda). \tag{10.14}$$

Let us first calculate the principal symbol of the PDO

$$(J^+)^* T^- - (J_0^+)^* T_0^- = (J^+ - J_0^+)^* T^- + (J_0^+)^*(T^- - T_0^-). \tag{10.15}$$

Below we use formulas (see subsection 4.3) for the principal symbol of the adjoint of a PDO and of the product of PDO. Recall that the phases $\Phi_0^\pm$ were defined by (10.4) and $\zeta^\pm(x,\xi) = \eta(x)\sigma^\pm(\langle \hat{x}, \hat{\xi}\rangle)$.

It follows from (10.13) that the principal symbol of the PDO $J^+ - J_0^+$ is

$$i2^{-1} \exp(i\Phi_0^+(x,\xi)) \int_0^\infty V(x+s\xi)ds\, \zeta^+(x,\xi)\psi(|\xi|^2).$$

The symbol of PDO $T^\pm$ was calculated in Section 9. It is given by formula (4.29) with $\tau_s^\pm$ defined by (9.22) and $\tau_r^\pm$ satisfying (9.23). The symbol of PDO $T_0^\pm$ is given by the same relations with $a^\pm$ replaced by $a_0^\pm$. In particular, the principal symbol of PDO T^- is

$$-2i \exp(i\Phi_0^-(x,\xi))\langle \xi, \nabla\zeta^-(x,\xi)\rangle \psi(|\xi|^2), \quad \nabla = \nabla_x,$$

(cf. (4.30)). Thus the principal symbol of $(J^+ - J_0^+)^* T^-$ is

$$-\exp(i\Phi_0(x,\xi)) \int_0^\infty V(x+s\xi)ds\, \langle \xi, \nabla\zeta^-(x,\xi)\rangle \psi^2(|\xi|^2),$$

where

$$\Phi_0(x,\xi) = \Phi_0^-(x,\xi) - \Phi_0^+(x,\xi) = 2^{-1}\int_{-\infty}^\infty (V_0(s\xi) - V_0(x+s\xi))ds.$$

Further, comparing expressions for symbols of PDO T_0^- and T^-, we find that the principal symbol of $T^- - T_0^-$ is

$$-\exp(i\Phi_0^-(x,\xi))(a^-(x,\xi) - a_0^-(x,\xi))\langle \xi, \nabla\zeta^-(x,\xi)\rangle \psi(|\xi|^2).$$

Replacing here $a^- - a_0^-$ by the right-hand side of (10.13), we see that the principal symbol of $(J_0^+)^*(T^- - T_0^-)$ is

$$-\exp(i\Phi_0(x,\xi)) \int_{-\infty}^0 V(x+s\xi)ds\, \langle \xi, \nabla\zeta^-(x,\xi)\rangle \psi^2(|\xi|^2).$$

Let us now put together the expressions obtained.

Lemma 10.7 *The principal symbol of PDO* (10.15) *equals*

$$-\exp(i\Phi_0(x,\xi))\int_{-\infty}^{\infty} V(x+s\xi)dt\,\langle\xi,\nabla\zeta^-(x,\xi)\rangle\psi^2(|\xi|^2).$$

Recall that the principal symbol of PDO $\Gamma_0(\lambda)A\Gamma_0^*(\lambda)$ is constructed from that of a PDO A by formula (7.20). Therefore (cf. calculation of integral (9.12)) Lemma 10.7 implies

Lemma 10.8 *The principal symbol of PDO* (10.14) *acting in the space* $L_2(\mathbb{S}^{d-1})$ *equals*

$$(4\pi k)^{-1}\exp(ik^{-1}\Phi_0(k^{-1}b,\omega))\int_{-\infty}^{\infty} V(k^{-1}b+\tau\omega)d\tau, \tag{10.16}$$

where $|\omega|=1$, $\langle b,\omega\rangle=0$.

Let us return to scattering matrices $S^\pm(\lambda)$. It follows from (10.10) that

$$S^+=I+(S_1-S_0)S_0^*,\quad S^-=I+S_0^*(S_1-S_0).$$

According to Theorem 9.10, the principal symbol of the operator S_0 equals

$$\exp(ik^{-1}\Phi_0(k^{-1}b,\omega)).$$

Comparing this expression with (10.16), we see that the terms containing Φ_0 cancel each other in the principal symbol of the operator $S^\pm - I$.

Theorem 10.9 *Let functions* V_0 *and* V *satisfy, for all* κ, *assumptions* (10.5) *where* $\rho_0>1/2$ *and* (4.1) *where* $\rho>1$, *respectively. Then the principal symbol of the PDO* $-(2\pi i)^{-1}(S^\pm(\lambda)-I)$ *is given by equality* (8.13).

This is exactly the same result as in the case $V_0=0$. We emphasize that both operators $S^+(\lambda)$ and $S^-(\lambda)$ have the same principal symbol. Theorem 10.9 remains true under assumptions of subsection 4.5.

Recall that the scattering matrix $S(\lambda;H,H_0)$ coincides with the operators $S^+(\lambda)$ and $S^-(\lambda)$ up to a unitary equivalence. Let us write eigenvalues $\mu_n^\pm(\lambda)$ of $S(\lambda;H,H_0)$ in the form (8.7). The following conclusion can be deduced from Theorem 10.9 quite similarly to Theorem 8.4.

Corollary 10.10 *In addition to the assumptions of Theorem* 10.9, *suppose that condition* (8.10) *holds. Then the phases* $\phi_n^\pm$ *have asymptotics* (8.11) *with the coefficient* $\mathcal{V}_\pm$ *defined by* (8.8), (8.9).

Part III

The multiparticle Schrödinger operator and related problems

Our presentation in this part is concentrated around the Schrödinger operator of N interacting particles. As was already mentioned in Introduction, the scattering problem becomes multichannel if $N \geq 3$. The complete classification of all possible channels is called asymptotic completeness; see Section 11 for the precise formulation of this result. For example, for $N = 3$ asymptotic completeness means that either all three particles form a bound state or, asymptotically, two particles are in a bound state and the third is free or all three particles are free. We give sketches both of the "old" Faddeev approach to three-particle scattering relying on a non-standard perturbation theory (Section 12) and of a "new" one using the limiting absorption principle and radiation estimates (Section 13). Both these methods have its advantages and drawbacks. Stationary representations of wave operators and scattering matrices are discussed in Section 14.

The multiparticle problem with long-range pair potentials is out of the scope of this survey (see the book [41]). However, in Section 15 we show that if pair potentials decay slower than $|x|^{-1/2}$, then the traditional picture of scattering breaks down. Actually, we construct an asymptotic evolution of a three-particle system which is intermediary between two types of scattering channels discussed above. For such evolution, the bound state of a couple of particles depends on a position of the third particle, and it is destroyed asymptotically. This scattering channel corresponds to a new class of asymptotic solutions of the time-dependent Schrödinger equation.

The discrete analogue of the N-particle Schrödinger operator is discussed in Section 16. Here additional problems appear.

Finally, scattering of waves (described, say, by the operator $-\Delta$) on unbounded obstacles is considered in Section 17. We present two different approaches to solution of this problem for the Dirichlet boundary condition. The case of other boundary conditions (for example, Neumann) remains open.

11. SETTING THE SCATTERING PROBLEM

11.1. Definitions of Hamiltonians. Let us recall the definition of N-particle

Schrödinger operator (Hamiltonian) $\mathbf{H} = \mathbf{H}_0 + \mathbf{V}$. If dimensions of particles are equal to k, then the configuration space of the system is $L_2(\mathbb{R}^{kN})$. The operator of kinetic energy (the "unperturbed" Hamiltonian) is

$$\mathbf{H}_0 = -\sum_{j=1}^{N} (2m_j)^{-1} \Delta_{x_j}, \tag{11.1}$$

where m_j are masses of particles. The operator of potential energy of pair interactions of particles (the perturbation) $\mathbf{V}$ is the operator of multiplication by the function

$$V(x) = \sum_{i<j} V^{(ij)}(x_i - x_j), \quad i,j = 1, \ldots, N, \tag{11.2}$$

where the functions $V^{(ij)}(y)$ tend to zero as $|y| \to \infty$ in $\mathbb{R}^k$. However the function $V(x) \not\to 0$ as $|x| \to \infty$ in $\mathbb{R}^{kN}$ if at least one of the distances $|x_i - x_j|$ between particles remains bounded. This difficulty is manifest even for two particles ($N = 2$), but it disappears if the motion of the center of mass of the system is removed.

This means the following. Let the subspace X^{cm} of $\mathbb{R}^{kN}$ be distinguished by the condition

$$\sum_{j=1}^{N} m_j x_j = 0, \tag{11.3}$$

and let X_{cm} be the orthogonal complement to X^{cm} in the space $\mathbb{R}^{kN}$ endowed with the scalar product

$$\sum_{j=1}^{N} m_j \langle x_j, y_j \rangle. \tag{11.4}$$

Then $L_2(\mathbb{R}^{kN}) = L_2(X_{cm}) \otimes L_2(X^{cm})$ and

$$\mathbf{H} = K \otimes I + I \otimes H, \quad K = -(2M)^{-1} \Delta_{\mathbf{x}}, \quad \mathbf{x} = M^{-1} \sum_{j=1}^{N} m_j x_j, \quad M = \sum_{j=1}^{N} m_j, \tag{11.5}$$

where K is the kinetic energy operator of the center of mass motion. The operator $H = H_0 + V$ acts in the space $L_2(X^{cm})$, the precise form of the differential operator H_0 depends on the choice of coordinates in X^{cm}, and V is again multiplication by function (11.2). If $N = 2$, then $V(x) \to 0$ as $|x| \to \infty$, $x \in X^{cm}$, but this is no longer true for $N \geq 3$. According to (11.5) the spectral and scattering theories for the operator $\mathbf{H}$ reduce to those for the operator H. However, for $N \geq 3$, this reduction is not really helpful.

It is convenient for us to study a more general class of operators introduced by S. Agmon [3]. It allows us, for example, to include automatically three (and more) particle interactions. More important, consideration of this class of Hamiltonians unravels better the intrinsic geometry of the problem (and permits to avoid a use of complicated coordinate systems) although an intuitive picture of moving classical particles is to some degree lost. The generalized N-particle Hamiltonian is the (self-adjoint) Schrödinger operator $H = -\Delta + V(x)$ on the Hilbert space $\mathcal{H} = L_2(\mathbb{R}^d)$

whose potential energy $V(x)$ has the following special structure. Suppose that some finite number α_0 of subspaces X^α of $X := \mathbf{R}^d$ are given, and let x^α, x_α be the orthogonal projections of $x \in X$ on X^α and $X_\alpha = X \ominus X^\alpha$, respectively. We assume that

$$V(x) = \sum_{\alpha=1}^{\alpha_0} V^\alpha(x^\alpha), \tag{11.6}$$

where V^α is a real function of the variable x^α which tends to 0 as $|x^\alpha| \to \infty$. Thus $V^\alpha(x^\alpha) \to 0$ as $|x| \to \infty$ off any conical neighbourhood of X_α, but $V^\alpha(x^\alpha)$ is constant on planes parallel to X_α. Due to this property the structure of the spectrum of H is much more complicated than in the two-particle case.

Clearly, operator (11.1) reduces to $-\Delta$ after the change $(2m_j)^{1/2}x_j = x_j'$ and potential energy (11.2) is a particular case of (11.6) if $\alpha = (ij)$, $i < j$, and

$$x^\alpha = (m_i + m_j)^{-1/2}(m_j^{1/2}x_i' - m_i^{1/2}x_j').$$

The two-particle Hamiltonian H with fixed center of mass is recovered if $\alpha_0 = 1$ and $X^1 = X$ in (11.6). The three-particle problem (again with a motion of center of mass removed) is distinguished from the general situation by the condition that $X_\alpha \cap X_\beta = \{0\}$ for $\alpha \neq \beta$. We emphasize that the generalized N-particle Hamiltonian includes the case where some of the masses m_j in (11.1) are infinite.

11.2. Cluster decompositions. The spectral theory of the operator H starts with the following geometrical construction. Let us consider linear sums

$$X^a = X^{\alpha_1} + X^{\alpha_2} + \ldots + X^{\alpha_k} \tag{11.7}$$

of the subspaces X^{α_j}. Orthogonal complements

$$X_a = X \ominus X^a = X_{\alpha_1} \cap X_{\alpha_2} \cap \ldots \cap X_{\alpha_k}$$

of X^a are usually called collision planes. Since $X = X_a \oplus X^a$, the space $\mathcal{H}$ splits into the tensor product

$$L_2(X) = L_2(X_a) \otimes L_2(X^a). \tag{11.8}$$

Below x^a and x_a are the orthogonal projections of $x \in X$ on the subspaces X^a and X_a, respectively.

For each a, define an auxiliary operator $H_a = -\Delta + V^a$ with a potential

$$V^a = \sum_{X^\alpha \subset X^a} V^\alpha \tag{11.9}$$

which does not depend on x_a. Set

$$K_a = -\Delta_{x_a}, \quad H^a = -\Delta_{x^a} + V^a \quad \text{and} \quad H_a = K_a \otimes I + I \otimes H^a, \tag{11.10}$$

where the tensor product is the same as in (11.8).

Let X^{a_0} be the largest of all subspaces (11.7). If $X^{a_0} \neq X$, then (11.10) makes sense for $a = a_0$ and $H_{a_0} = H$. Therefore scattering theory for the operator H

reduces to that for the operator H^{a_0}. Thus, without loss of generality, we may suppose that $X = X^{a_0}$, but this assumption is not really necessary and we do not make it. In terms of N-particle systems, $X^{a_0} = X^{cm}$ and the passage from H to H^{a_0} corresponds to removal of center of mass, that is to the passage from $\mathbf{H}$ to H (see (11.5)).

We denote by $\mathcal{X}$ the set of all subspaces X^a with $X^0 := \{0\} \in \mathcal{X}$ included in it but X excluded. The set of collision planes X_a for $X^a \in \mathcal{X}$ will be denoted by $\mathcal{X}'$.

It is useful to keep in mind the following model Hamiltonian

$$H = -\Delta + V^1(x^1) + V^2(x^2) + V^3(x^3), \quad x \in \mathbf{R}^3, \tag{11.11}$$

where x^α are orthogonal projections on lines $X^\alpha \subset \mathbf{R}^3$. The operator (11.11) describes 4 one-dimensional particles with only 3 non-trivial pair interactions (say, $V^{(12)}, V^{(23)}, V^{(34)}$ in notation (11.2)). A potential $V^\alpha(x^\alpha)$ does not tend to zero in a conical neighbourhood of the plane $X_\alpha = \mathbf{R}^3 \ominus X^\alpha$. Two potentials $V^\alpha(x^\alpha)$ and $V^\beta(x^\beta)$ do not tend to zero in a conical neighbourhood of the line $X_{\alpha\beta} = X_\alpha \cap X_\beta$. The set $\mathcal{X}'$ consists of the space $X = \mathbf{R}^3$, the planes X_1, X_2, X_3 and the lines X_{12}, X_{23}, X_{31} (see Figure 11.1).

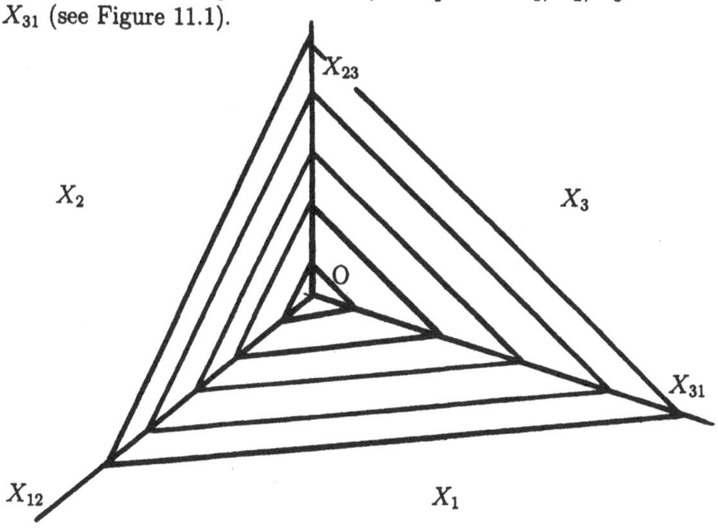

Figure 11.1. The configuration space of the Hamiltonian (11.11).

In the multiparticle terminology, $a = \{C_1, \dots, C_p\}$ parametrizes decompositions of an N-particle system into noninteracting clusters C_l, that is

$$C_1 \cup \dots \cup C_p = \{1, \dots, N\}, \quad C_l \cap C_k = \emptyset \quad \text{if} \quad l \neq k, \quad p \geq 2. \tag{11.12}$$

We denote by $\kappa_l = |C_l|$ the number of particles in a cluster C_l. The operator

$$\mathbf{H}_a = \mathbf{H}_0 + \mathbf{V}^a, \quad \mathbf{V}^a = \sum_{l=1}^{p} \sum_{i,j \in C_l, i<j} V^{(ij)} \tag{11.13}$$

is the Hamiltonian of the N-particle system with interactions between different clusters neglected. The "external" variable

$$x_a = (\mathbf{x}_1, \mathbf{x}_2, \dots, \mathbf{x}_p), \quad \mathbf{x}_l = M_l^{-1} \sum_{j \in C_l} m_j x_j, \quad M_l = \sum_{j \in C_l} m_j, \tag{11.14}$$

describes positions of centers of mass of the clusters. The "internal" variable x^a is the set of all numbers $x_j - \mathbf{x}_l$ for all $j \in C_l$ and all $l = 1, \dots, p$. Of course, for each l only $\kappa_l - 1$ of variables $x_j - \mathbf{x}_l$ are independent. Clearly, the subspace X^a is determined by the condition

$$\sum_{j \in C_l} m_j x_j = 0, \quad l = 1, \dots, p,$$

and X_a is the orthogonal complement to X^a with respect to the scalar product (11.4). Of course, the same procedure applies to the operator H considered in the space $L_2(X^{cm})$ (recall that $X^{cm} \subset \mathbf{R}^{kN}$ was distinguished by condition (11.3)). In particular, the cluster Hamiltonian $H_a = H_0 + V^a$ where V^a is defined by the same formula (11.13) as $\mathbf{V}^a$ but V^a acts in $L_2(X^{cm})$.

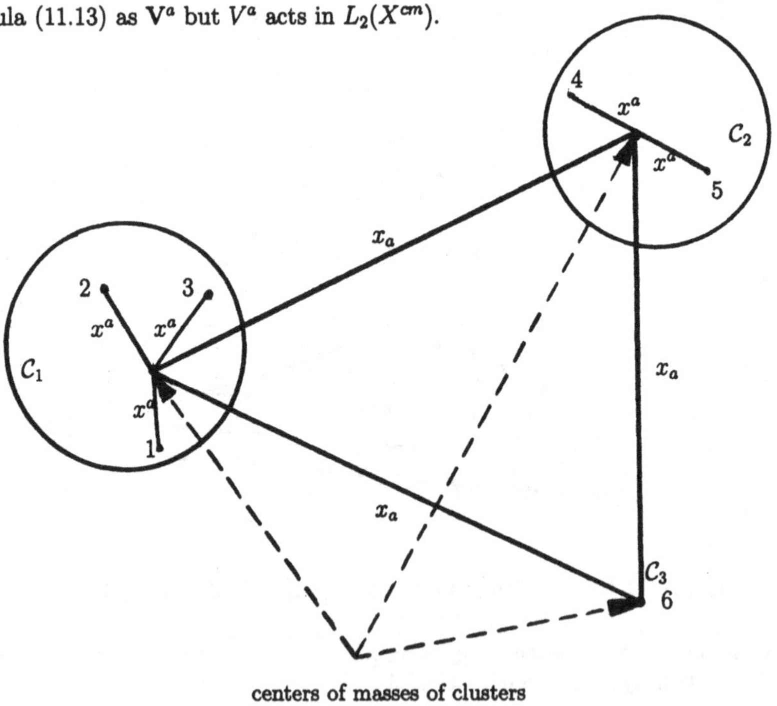

Figure 11.2. A decomposition of 6 particles in 3 clusters: $C_1 = (1, 2, 3)$, $C_2 = (4, 5)$, $C_3 = (6)$.

11.3. Asymptotic completeness. Below the index a labels all subspaces $X^a \in \mathcal{X}$. Let P^a be the orthogonal projection in $L_2(X^a)$ on the subspace $\mathcal{H}_{H^a}^{(p)}$

spanned by all eigenvectors $\psi^{a,n}$ of H^a, and let $\mathsf{P}_a = I \otimes \mathsf{P}^a$ where the tensor product is defined by (11.8). Then P_a commutes with the operator H_a. Set also $K_0 = H_0 = -\Delta, \mathsf{P}_0 = I$. The fundamental result of scattering theory for generalized N-particle Schrödinger operators with short-range potentials V^α is the following

Theorem 11.1 *Let $H = -\Delta + V(x)$ with potential* (11.6). *Suppose that for all α*

$$|V^\alpha(x^\alpha)| \leq C(1 + |x^\alpha|)^{-\rho}, \quad \rho > 1. \tag{11.15}$$

Then, for all a, the wave operators

$$W_a^\pm = W^\pm(H, H_a; \mathsf{P}_a) \tag{11.16}$$

exist and are isometric on the ranges $\operatorname{Ran} \mathsf{P}_a$ *of projections* P_a. *The subspaces* $\operatorname{Ran} W_a^\pm$ *are mutually orthogonal, and scattering is asymptotically complete:*

$$\bigoplus_a \operatorname{Ran} W_a^\pm = \mathcal{H}^{(ac)}, \quad \mathcal{H}^{(ac)} = \mathcal{H}_H^{(ac)}. \tag{11.17}$$

We note (see Theorem 13.1 below) that the singular continuous spectrum of H is empty, so $\mathcal{H}^{(ac)}$ can be replaced by $\mathcal{H} \ominus \mathcal{H}^{(p)}$ where $\mathcal{H}^{(p)}$ is spanned by all eigenvectors of H. This remark is of course non-trivial in the case $X^{a_0} = X$ only. Otherwise $H = H_{a_0}$ is absolutely continuous as follows from (11.10) and

Lemma 11.2 *Let H_1 and H_2 be self-adjoint operators in Hilbert spaces $\mathcal{H}_1$ and $\mathcal{H}_2$, respectively. Then the operator*

$$H = H_1 \otimes I + I \otimes H_2 \quad \text{in} \quad \mathcal{H}_1 \otimes \mathcal{H}_2$$

is absolutely continuous if (at least) one of these operators is absolutely continuous.

Indeed, let $f = f_1 \otimes f_2$, $N \subset \mathbb{R}$ be a Borel set and $N_\lambda = N - \lambda$. Then

$$(E(N)f, f) = \int_{-\infty}^{\infty} (E_1(N_\lambda)f_1, f_1) d(E_2(\lambda)f_2, f_2).$$

If, for example, H_1 is absolutely continuous and $|N| = 0$, then $|N_\lambda| = 0$ and $(E_1(N_\lambda)f_1, f_1) = 0$ for all $\lambda \in \mathbb{R}$.

The result of Theorem 11.1 (asymptotic completeness) can be reformulated in terms of scattering theory in a couple of spaces. Eigenvectors $\psi^{a,n}$ are supposed to be normalized and orthogonal if the corresponding eigenvalues $\lambda^{a,n}$ coincide. Let us write $\mathbf{a}$ instead of a couple $\{a, n\}$ and introduce an auxiliary space

$$\hat{\mathcal{H}} = \bigoplus_{\mathbf{a}} \mathcal{H}_{\mathbf{a}}, \quad \mathcal{H}_{\mathbf{a}} = \mathcal{H}_a = L_2(X_a), \tag{11.18}$$

and an auxiliary operator

$$\hat{H} = \bigoplus_{\mathbf{a}} K_{\mathbf{a}}, \quad K_{\mathbf{a}} = K_a + \lambda^{\mathbf{a}}, \tag{11.19}$$

in this space. Here and below the sums are taken over all $\mathbf{a}$. We define an identification $\hat{J} : \hat{\mathcal{H}} \to \mathcal{H}$ by the relations

$$\hat{J} = \sum_{\mathbf{a}} J^{\mathbf{a}}, \quad J^{\mathbf{a}} f_{\mathbf{a}} = f_{\mathbf{a}} \otimes \psi^{\mathbf{a}}, \tag{11.20}$$

where the tensor product is determined by the decomposition (11.8). In particular, $\lambda^0 = 0$ and $J^0 = I$. It follows from (11.10) that

$$H_a J^{\mathbf{a}} = J^{\mathbf{a}} K_{\mathbf{a}}$$

and hence Theorem 11.1 may be reformulated as follows.

Theorem 11.1 bis *Under assumption* (11.15) *the wave operators* $W^{\pm}(H, \hat{H}; \hat{J})$ *exist, are isometric and complete.*

Theorem 11.1 shows that, for states orthogonal to eigenvectors of H, evolution of an N-particle system decomposes asymptotically into a sum of evolutions which are "free" in external variables x_a and are determined by eigenvalues and eigenfunctions of the Hamiltonians H^a in internal variables x^a. This is formulated in the following

Corollary 11.3 *For any* $f \in \mathcal{H}^{(ac)}$ *there exist vectors* $f_{\mathbf{a}}^{\pm} \in \mathcal{H}_a$

$$(\text{actually,} \quad f_{\mathbf{a}}^{\pm} = W^{\pm}(H, K_{\mathbf{a}}; J^{\mathbf{a}})^* f, \quad \mathbf{a} = \{a, n\})$$

such that

$$\exp(-iHt)f = \sum_{\mathbf{a}} \exp(-iK_{\mathbf{a}}t) f_{\mathbf{a}}^{\pm} \otimes \psi^{\mathbf{a}} + o(1), \quad t \to \pm\infty, \tag{11.21}$$

where, for every a, *the tensor product is the same as in* (11.8) *and the term* $o(1)$ *tends to zero in* $L_2(X)$.

Corollary 11.4 *For all* a, *the wave operators* $W^{\pm}(H, H_a)$ *exist.*

Indeed, applying (11.21) to the function $\exp(-iH_a t)f$, we see that

$$\exp(-iH_a t)f = \sum_{X^b \subset X^a} \sum_n \exp(-i(K_b + \lambda^{b,n})t) f_{b,n}^{\pm} \otimes \psi^{b,n} + o(1), \quad t \to \pm\infty.$$

Since, by Theorem 11.1, all wave operators $W^{\pm}(H, H_b; \mathsf{P}_b)$ exist, this implies the existence of $W^{\pm}(H, H_a)$.

In the multiparticle terminology, the wave operator $W^{\pm}(H, K_{\mathbf{a}}; J^{\mathbf{a}})$ describes the scattering channel where a system of N interacting particles splits up asymptotically (for $t \to \pm\infty$) into non-interacting clusters $C_1, \ldots, C_p$, $p \geq 2$, and particles from the same cluster C_l are in the bound state (if there are more than one particle in C_l) given by the function $\psi^{\mathbf{a}}(x^a)$.

11.4. Comments. In the three-particle case Theorem 11.1 was first obtained, under some additional assumptions, by L. D. Faddeev [48] (see also [59], [149]), who used a set of equations he derived for the resolvent of H. Faddeev's approach will be briefly outlined in the next section. The optimal formulation in the three particle case is due to V. Enss [44] who used a purely time-dependent method.

Another approach to asymptotic completeness relies on the Mourre estimate (see subsection 13.1) which gives an a priori estimate for the resolvent of the N-particle Hamiltonian (the limiting absorption principle). This approach is intrinsically non-perturbative and goes back to I. Sigal and A. Soffer [140]. In Section 13 we give rather a detailed sketch of the proof of asymptotic completeness based on the paper [166]. To a some extent, it is resemblent to the earlier proof of G. M. Graf [60] which can be found in the book [41]. Note also that a proof of asymptotic completeness, in some sense intermediary between those of [60] and [166], was given in [68]. In contrast to [60] we fit N-particle scattering theory into the standard framework of the smooth perturbation theory. Its advantage is that it admits two equivalent formulations, time-dependent and stationary, whereas Graf's method is intrinsically time-dependent. This allows us [167, 168] to obtain stationary (in terms of resolvents) formulas for the basic objects of the theory: wave operators, scattering matrix (see Section 14), etc. The approaches of [48] and of [140, 60, 166] are quite different, and at the moment there is no bridge between them.

We note also the paper [114] by R. Lavine where, prior to the Mourre estimate, the asymptotic completeness has been proven for an arbitrary number of particles with repulsive potentials. In this paper the commutator method of [133, 99] was first applied to N-particle Schrödinger operators.

By virtue of relations (3.3), (3.4) valid for every operator K_a, asymptotics (11.21) can be rewritten as

$$\exp(-iHt)f = \sum_{\mathbf{a}} \exp(i\Phi_{\mathbf{a}}(x_a,t))(2it)^{-d_a/2}\hat{f}_{\mathbf{a}}^{\pm}(x_a/(2t))\psi^{\mathbf{a}}(x^a) + o(1), \qquad (11.22)$$

where $t \to \pm\infty$, $d_a = \dim X_a$, $\Phi_{\mathbf{a}}(x_a,t) = x_a^2(4t)^{-1} - \lambda^{\mathbf{a}}t$ and $\hat{f}_{\mathbf{a}}^{\pm}$ is the Fourier transform of $f_{\mathbf{a}}^{\pm}$. For N-particle systems with long-range pair potentials V^α the result is almost the same if condition (4.1) with some $\rho > \sqrt{3}-1$ is fulfilled for all functions $V^\alpha(x^\alpha)$. In this case again every $f \in \mathcal{H}^{(ac)}$ satisfies (11.22) with functions

$$\Phi_{a,n}(x_a,t) = x_a^2(4t)^{-1} - \lambda^{a,n}t - t\int_0^1 V_a(sx_a)ds,$$

where $V_a(x_a) = V(x_a,0) - V^a(0)$. This result (asymptotic completeness) was obtained by V. Enss [45, 46] for $N = 3$ and extended by J. Dereziński [39] to an arbitrary number of particles (see also [142, 143]). Dereziński's method is different from [45, 46] and uses some ideas of I. Sigal and A. Soffer [141]. In particular, it relies on a property of asymptotic clustering which, roughly speaking, means that for large t a system decomposes into independently moving clusters of particles. The relative motion of the centers of masses of the clusters is influenced by long-range intercluster potentials. Asymptotic clustering is weaker than asymptotic completeness since it does not assert that the clusters are in bound states of their internal Hamiltonians. For arbitrary $\rho > 1/2$, asymptotic clustering was verified in [40]. Dereziński's proof is exposed in the book [41].

11.5. Dispersive Hamiltonians. Let us finally consider a more general case where the kinetic energy $|\xi|^2$ of a particle is replaced by a sufficiently arbitrary function $\omega(\xi)$. As typical examples, we note $\omega(\xi) = (|\xi|^2 + m^2)^{1/2}$ or $\omega(\xi) = |\xi|^4$. Such Hamiltonians are called dispersive. Thus we now suppose that $\mathbf{H} = \mathbf{H}_0 + \mathbf{V}$, where $\mathbf{H}_0$ acts in the momentum representation as multiplication by the function

$$h(\xi) = \sum_{j=1}^{N} \omega(\xi_j), \quad \xi_j \in \mathbf{R}^k,$$

and, as before, $\mathbf{V}$ is given by formula (11.2). The operator $\mathbf{H}$ acts in the space $\mathcal{H} = L_2(\mathbf{R}^{kN})$, and its center of mass motion cannot be separated.

Let us again consider an arbitrary cluster decomposition (11.12), and let the potential energy $\mathbf{V}^a$ of interaction of particles belonging to the same cluster be determined by formula (11.13). Define the internal x^a and external x_a (see formula (11.14)) variables as in subsection 11.2 with $m_j = 1$, $M_l = \kappa_l$; then the corresponding subspaces X^a and X_a are orthogonal in $L_2(\mathbf{R}^{kN})$. Let ξ^a and ξ_a be the internal and external momentum variables. Then the cluster Hamiltonian $\mathbf{H}_a = \mathbf{H}_0 + \mathbf{V}_a$ acts in the space $\mathcal{H}_a = L_2(X_a) \otimes L_2(X^a)$ as multiplication by the operator-function

$$\mathbf{H}^a(\xi_a) = h(\xi_a, \xi^a) + \mathbf{V}^a(x^a) : L_2(X^a) \to L_2(X^a)$$

depending on the external momentum $\xi_a \in X_a$. As always, we do not distinguish in notation the configuration and momentum spaces. Moreover, to simplify presentation in this subsection we use the same notation for different representations of the same operator. Clearly, $h(\xi_a, \xi^a) = |\xi^a|^2 + |\xi_a|^2$ if $\omega(\xi_j) = |\xi_j|^2$, but in the general case the variable ξ_a cannot be separated. This construction makes sense also for $p = 1$. Then $C_1 = \{1, \ldots, N\}$, the subspace X^{a_0} is determined by the condition

$$x_1 + x_2 + \cdots + x_N = 0,$$

$\mathbf{V}^{a_0} = \mathbf{V}$ and $\mathbf{H}$ acts as multiplication by the operator-function $\mathbf{H}^{a_0}(\xi_{a_0})$ where $\xi_{a_0} \in \mathbf{R}^k$ is the total momentum of the system. Thus for a dispersive Hamiltonian, the separation of center of mass motion is formulated in terms of a diagonal for $\mathbf{H}$ direct integral.

It is posible that, at least for some values of ξ_a, the operators $\mathbf{H}^a(\xi_a)$ have point spectrum. Let $\lambda^{\mathbf{a}}(\xi_a)$, $\mathbf{a} = \{a, n\}$, $n = 1, 2, \ldots$, and $\psi^{\mathbf{a}}(x^a, \xi_a)$ be an eigenvalue and the corresponding normalized eigenfunction of the operator $\mathbf{H}^a(\xi_a)$ for ξ_a belonging to some set $U_{\mathbf{a}} \subset X_a$. Denote by $\mathcal{H}_{\mathbf{a}}^{(bs)}$ the subspace of all functions $\psi^{\mathbf{a}}(x^a, \xi_a) f_{\mathbf{a}}(\xi_a)$ where $f_{\mathbf{a}} \in L_2(U_{\mathbf{a}})$ and

$$\mathcal{H}_a^{(bs)} = \bigoplus_n \mathcal{H}_{a,n}^{(bs)}.$$

In particular, vectors from the subspace $\mathcal{H}^{(bs)} = \mathcal{H}_{a_0}^{(bs)}$ play the role of bound states of the whole system. The operator $\mathbf{H}_a$ reduces on $\mathcal{H}_{\mathbf{a}}^{(bs)}$ to multiplication by $\lambda^{\mathbf{a}}(\xi_a)$. Recall (see Chapter 16, v.4 of [134]) that, apart from the absolutely continuous component, an operator of multiplication may have only eigenvalues of infinite multiplicity (if some $\lambda^{\mathbf{a}}(\xi_a) = const$ on a set of positive measure). Therefore, the operator

$\mathbf{H}_a$ does not have the singular continuous spectrum on the subspace $\mathcal{H}_{\mathbf{a}}^{(bs)}$. Probably (recall that $\lambda^{\mathbf{a}}(\xi_a) = \lambda^{\mathbf{a}} + |\xi_a|^2$ if $h(\xi) = |\xi|^2$), the functions $\lambda^{\mathbf{a}}(\xi_a)$ cannot take constant values on sets of positive measure. So we expect that the operators $\mathbf{H}_a$ on the subspaces $\mathcal{H}_{\mathbf{a}}^{(bs)}$ and, actually, on the whole space $\mathcal{H}$ are absolutely continuous.

Let P_a be the orthogonal projection on the subspace $\mathcal{H}_{\mathbf{a}}^{(bs)}$. Then a natural formulation of scattering theory for the Hamiltonian $\mathbf{H}$ is almost the same as that given by Theorem 11.1. For all cluster decompositions (11.12), the wave operators $W^{\pm}(\mathbf{H}, \mathbf{H}_a; \mathsf{P}_a)$ exist, their ranges are mutually orthogonal and the asymptotic completeness

$$\bigoplus_a \operatorname{Ran} W^{\pm}(\mathbf{H}, \mathbf{H}_a; \mathsf{P}_a) = \mathcal{H} \ominus \mathcal{H}^{(bs)} \tag{11.23}$$

holds. It is easy to see that, in the case $h(\xi) = |\xi|^2$, equality (11.23) reduces to (11.17). Given that the eigenvalues $\lambda^{\mathbf{a}_0}(\xi_{a_0})$ of the operators $\mathbf{H}^{a_0}(\xi_{a_0})$ cannot be constants on sets of positive measure, equality (11.23) implies that the operator $\mathbf{H}$ is absolutely continuous. The last result is of course trivial (see Lemma 11.2) in the case $h(\xi) = |\xi|^2$. Thus we finish with

Problem 11.5 *Prove the above hypotheses and, in particular, equality* (11.23) *for sufficiently general functions* ω.

A short discussion of its possible solution can be found in subsection 13.4.

12. RESOLVENT EQUATIONS FOR THREE-PARTICLE SYSTEMS

In this section we describe Faddeev's approach [48] to three-particle scattering.

12.1. Faddeev's equations. Let the Hamiltonian **H** be defined by formulas (11.1), (11.2) where $N = 3$ and the dimension of particles $k \geq 3$. It is now important that the center of mass motion be removed. Then the operator H acts in the space $\mathcal{H} = L_2(X)$ where the subspace $X \subset \mathbf{R}^{3k}$ is distinguished by condition (11.3). A natural choice of coordinates in X is given by one of the three sets of Jacobi variables:

$$x^{12} = x_1 - x_2, \quad x_{12} = x_3 - (m_1 + m_2)^{-1}(m_1 x_1 + m_2 x_2) \tag{12.1}$$

and similarly for x^{23}, x_{23} and x^{31}, x_{31}. In coordinates x^α, x_α the operator of kinetic energy is determined by the formula

$$H_0 = -(2m_\alpha)^{-1}\Delta_{x_\alpha} - (2m^\alpha)^{-1}\Delta_{x^\alpha},$$

where, for example,

$$(m^{12})^{-1} = m_1^{-1} + m_2^{-1}, \quad m_{12}^{-1} = (m_1 + m_2)^{-1} + m_3^{-1}.$$

In the three-particle case the resolvent equation (1.13) is not Fredholm even for $\operatorname{Im} z \neq 0$. To overcome this difficulty, Faddeev derived a system of equations for components of the resolvent. The entries of this system are constructed in terms of three Hamiltonians

$$H_\alpha = H_0 + V^\alpha, \tag{12.2}$$

containing only one pair interaction each. In this subsection it suffices to assume that condition (11.15) on pair potentials V^α is fulfilled for some $\rho > 0$ (or even that $V^\alpha(x^\alpha) \to 0$ as $|x^\alpha| \to \infty$). Let us write down the resolvent equation

$$R(z) = R_\alpha(z) - R_\alpha(z) \sum_{\beta \neq \alpha} V^\beta R(z), \quad R_\alpha(z) = (H_\alpha - z)^{-1}, \tag{12.3}$$

for each pair H_α, H. We denote by $\langle x^\alpha \rangle$ the multiplication by $(1 + |x^\alpha|^2)^{1/2}$ and by B^α the multiplication by a bounded function $V^\alpha(x^\alpha)(1 + |x^\alpha|^2)^l$, $l = \rho/2$, in the space $\mathcal{H}$. Multiplying (from the left) each of equations (12.3) by $\langle x^\alpha \rangle^{-l}$, we obtain a system (Faddeev's system) of three equations

$$\mathcal{R}_\alpha(z) = \langle x^\alpha \rangle^{-l} R_\alpha(z) - \sum_{\beta \neq \alpha} \langle x^\alpha \rangle^{-l} R_\alpha(z) \langle x^\beta \rangle^{-l} B^\beta \mathcal{R}_\beta(z) \tag{12.4}$$

for three operators $\mathcal{R}_\alpha(z) = \langle x^\alpha \rangle^{-l} R(z)$. Note that the resolvent $R(z)$ can be recovered from its components $\mathcal{R}_\alpha(z)$ by the formula

$$R(z) = R_0(z) - R_0(z) \sum_\alpha \langle x^\alpha \rangle^{-l} B^\alpha \mathcal{R}_\alpha(z).$$

Let us consider (12.4) as a vector equation

$$\mathbf{R}(z) = \mathbf{R}_0(z) - \mathbf{r}(z)\mathbf{R}(z) \tag{12.5}$$

for the operator $\mathbf{R}(z)$ with three components $\mathcal{R}_\alpha(z)$. Here $\mathbf{R}_0(z) = \{\langle x^\alpha\rangle^{-l} R_\alpha(z)\}$ and components of the "matrix" operator $\mathbf{r}(z)$ equal

$$\mathbf{r}_{\alpha,\alpha}(z) = 0, \quad \mathbf{r}_{\alpha,\beta}(z) = \langle x^\alpha\rangle^{-l} R_\alpha(z)\langle x^\beta\rangle^{-l} B^\beta. \tag{12.6}$$

Of course, $\mathbf{r}(z)$ is analytic in z for $\operatorname{Im} z \neq 0$. In contrast to (1.13), equation (12.5) is Fredholm.

Proposition 12.1 *Let condition* (11.15) *on pair potentials V^α be fulfilled for some $\rho > 0$. Then for $\operatorname{Im} z \neq 0$ the operator $\mathbf{r}(z)$ acting in the space $\mathfrak{H} = \mathcal{H} \oplus \mathcal{H} \oplus \mathcal{H}$ is compact and -1 is not its eigenvalue.*

Let us first check that $\mathbf{r}_{\alpha,\beta}(z) \in \mathfrak{S}_\infty$. Since

$$\langle x^\alpha\rangle^{-l} R_\alpha(z)\langle x^\beta\rangle^{-l} = (I - \langle x^\alpha\rangle^{-l} R_\alpha(z)\langle x^\alpha\rangle^{-l} B^\alpha)(\langle x^\alpha\rangle^{-l} R_0(z)\langle x^\beta\rangle^{-l}), \tag{12.7}$$

it suffices to verify that the operators $\langle x^\alpha\rangle^{-l} R_0(z)\langle x^\beta\rangle^{-l}$, $\alpha \neq \beta$, are compact. Approximating functions $\langle x^\alpha\rangle^{-l}$ by functions $q^\alpha(x^\alpha)$ with compact support, we reduce the problem to the inclusion $q^\alpha R_0(z) q^\beta \in \mathfrak{S}_\infty$. Recall that the integral kernel $r_0(x-y;z)$ of the free resolvent $R_0(z)$ decays exponentially as $|x-y| \to \infty$. Therefore the kernel

$$q^\alpha(x^\alpha) r_0(x-y;z) q^\beta(y^\beta)$$

of the operator $q^\alpha R_0(z) q^\beta$ belongs to the space $L_2(X \times X)$, and hence this operator belongs to the Hilbert-Schmidt class.

According to (12.6) the equation $f = -\mathbf{r}(z) f$ for $f = \{f_\alpha\}$ means that

$$f_\alpha = -\sum_{\beta \neq \alpha} \langle x^\alpha\rangle^{-l} R_\alpha(z)\langle x^\beta\rangle^{-l} B^\beta f_\beta. \tag{12.8}$$

In terms of the functions $u_\alpha = \langle x^\alpha\rangle^{-l} B^\alpha f_\alpha$ this system can be rewritten as

$$u_\alpha = -V^\alpha R_\alpha(z) \sum_{\beta \neq \alpha} u_\beta. \tag{12.9}$$

Now the resolvent identity for the pair H_0, H_α implies that

$$u_\alpha = -V^\alpha R_0(z) \sum_{\beta \neq \alpha} u_\beta + V^\alpha R_0(z) V^\alpha R_\alpha(z) \sum_{\beta \neq \alpha} u_\beta. \tag{12.10}$$

By (12.9) the second term in the right-hand side of (12.10) equals $-V^\alpha R_0(z) u_\alpha$ and hence

$$u_\alpha = -V^\alpha R_0(z) u, \quad \text{where} \quad u = \sum_\alpha u_\alpha. \tag{12.11}$$

Therefore $u = -V R_0(z) u$ and $\psi = R_0(z) u$ satisfies the equation $H\psi = z\psi$. Thus $\psi = 0$ which, by (12.11), implies that $u_\alpha = 0$. Finally, equation (12.8) shows that $f_\alpha = 0$ for all α.

12.2. The one-channel case. Our next goal is to study the limit of system (12.4) as $\operatorname{Im} z \to 0$. To that end we suppose that condition (11.15) on pair potentials is fulfilled for some $\rho > 2$. This implies, in particular, that every two-particle Hamiltonian

$$H^\alpha = -(2m^\alpha)^{-1}\Delta_{x^\alpha} + V^\alpha(x^\alpha)$$

acting in the space $L_2(\mathbf{R}^k)$ has only a finite set of negative eigenvalues $\lambda^{\alpha,n}$. Moreover, we assume that all three equations

$$f - 2m^\alpha\langle x^\alpha\rangle^{-l}\Delta^{-1}\langle x^\alpha\rangle^{-l}B^\alpha f = 0$$

have only trivial solutions in $L_2(\mathbf{R}^k)$. This means (see subsection 3.1) that operators H^α do not have zero-energy eigenvalues or resonances. By Proposition 3.1, under this assumption, the operator-functions

$$\langle x^\alpha\rangle^{-l}R^\alpha(z)\langle x^\alpha\rangle^{-l}, \quad R^\alpha(z) = (H^\alpha - z)^{-1}, \quad l > 1,$$

acting in the space $L_2(\mathbf{R}^k)$, are analytic in the complex plane cut along $[0,\infty)$, they have poles only at the points $\lambda^{\alpha,n}$ and are continuous up to the cut, the point $z = 0$ included.

Separating variables in the coordinates x^α, x_α defined by (12.1), we see that $H_\alpha = -(2m_\alpha)^{-1}\Delta_{x_\alpha} + H^\alpha$ (cf. (11.10)). In the "mixed" representation (ξ_α, x^α), where the Fourier transform in the variable x_α is performed and the variable ξ_α is dual to x_α, the operator $H_\alpha = -(2m_\alpha)^{-1}|\xi_\alpha|^2 + H^\alpha$. Therefore the resolvent $R_\alpha(z)$ of the operator H_α can be expressed via the kernel $r^\alpha(x^\alpha, y^\alpha; z)$ of the integral operator $R^\alpha(z)$ by the formula

$$(R_\alpha(z)g)(\xi_\alpha, x^\alpha) = \int_{\mathbf{R}^k} r^\alpha(x^\alpha, y^\alpha, z - (2m_\alpha)^{-1}|\xi_\alpha|^2)g(\xi_\alpha, y^\alpha)dy^\alpha. \tag{12.12}$$

It follows that

$$||\langle x^\alpha\rangle^{-l}R_\alpha(z)\langle x^\alpha\rangle^{-l}|| \leq \sup_{s\geq 0} ||\langle x^\alpha\rangle^{-l}R^\alpha(z-s)\langle x^\alpha\rangle^{-l}||. \tag{12.13}$$

In this subsection we assume that the operators H^α do not have negative eigenvalues. Then (12.13) implies that the operator-functions $\langle x^\alpha\rangle^{-l}R_\alpha(z)\langle x^\alpha\rangle^{-l}$, $l > 1$, remain bounded as the complex parameter z approaches $[0,\infty)$. It can me verified quite similarly that, actually, these functions are continuous in z. Thus we obtain

Proposition 12.2 *The operator-functions $\langle x^\alpha\rangle^{-l}R_\alpha(z)\langle x^\alpha\rangle^{-l}$, $l > 1$, are continuous in norm up to the cut along $[0,\infty)$. In particular, the operators $\langle x^\alpha\rangle^{-l}$, $l > 1$, are H_0-smooth.*

We need also the following "two-particle" result.

Lemma 12.3 *Let $v, w \in L_q(\mathbf{R}^k)$ for some $q \geq 2$. Then the operator-norm*

$$||v\exp(it\Delta)w||_{L_2(\mathbf{R}^k)} \leq |2t|^{-k/q}||v||_{L_q(\mathbf{R}^k)}||w||_{L_q(\mathbf{R}^k)}. \tag{12.14}$$

Indeed, the operator $\exp(it\Delta)$ is unitary in $L_2(\mathbf{R}^k)$ and according to (3.2)

$$\|\exp(it\Delta)f\|_{L_\infty(\mathbf{R}^k)} \leq |2t|^{-k/2}\|f\|_{L_1(\mathbf{R}^k)}.$$

Then the Riesz-Thorin interpolation theorem (see, e.g., [134], v.2) shows that

$$\|\exp(it\Delta)f\|_{L_{p'}(\mathbf{R}^k)} \leq |2t|^{-k(1/p-1/2)}\|f\|_{L_p(\mathbf{R}^k)}$$

for any $p \in [1,2]$ and $p^{-1}+(p')^{-1}=1$. Using twice the Hölder inequality, we obtain (12.14).

The next result is crucial for our study of the operator $\mathbf{r}(z)$.

Proposition 12.4 *The operator-functions $\langle x^\alpha\rangle^{-l}R_0(z)\langle x^\beta\rangle^{-l}$, $\alpha \neq \beta$, $l > 1$, are continuous in norm up to the cut along $[0,\infty)$.*

Indeed, by virtue of the representation

$$R_0(z) = i\int_0^\infty \exp(-itH_0)\exp(izt)dt, \quad \operatorname{Im} z > 0,$$

it suffices to check that

$$\|\langle x^\alpha\rangle^{-l}\exp(-itH_0)\langle x^\beta\rangle^{-l}\| = O(t^{-p}), \quad p > 1. \tag{12.15}$$

Suppose, for example, that $\alpha = (12)$, $\beta = (31)$. By the proof of (12.15), we may consider all operators in the space $L_2(\mathbf{R}^{3k})$, that is neglect separation of the center of mass motion. Taking into account that the operator Δ_{x_3} commutes with $\langle x^{12}\rangle^{-l}$ and Δ_{x_2} commutes with $\langle x^{31}\rangle^{-l}$, we see that (12.15) is equivalent to

$$\|\langle x^{12}\rangle^{-l}\exp(it(2m_1)^{-1}\Delta_{x_1})\langle x^{31}\rangle^{-l}\|_{L_2(\mathbf{R}^{3k})} = O(t^{-p}), \quad p > 1. \tag{12.16}$$

Let us apply now Lemma 12.3 in the variable x_1 for functions

$$v_2(x_1) = \langle x_1 - x_2\rangle^{-l} \quad \text{and} \quad v_3(x_1) = \langle x_1 - x_3\rangle^{-l}$$

depending on parameters x_2, x_3. Since they belong to the space $L_q(\mathbf{R}^k)$ for any $q > k/l$, we obtain the estimate

$$\|\langle x_1 - x_2\rangle^{-l}\exp(it(2m_1)^{-1}\Delta_{x_1})\langle x_1 - x_3\rangle^{-l}f\|^2_{L_2(\mathbf{R}^k)} \leq c|t|^{-2p}\|f\|^2_{L_2(\mathbf{R}^k)}, \tag{12.17}$$

where p is any number smaller than l and $p \leq k/2$. Therefore p can be chosen larger than 1 if $l > 1$ and $k \geq 3$. Integrating inequality (12.17) over x_2 and x_3, we get (12.16).

Combining identity (12.7) with Propositions 12.2 and 12.4, we arrive at

Proposition 12.5 *The operator-functions $\langle x^\alpha\rangle^{-l}R_\alpha(z)\langle x^\beta\rangle^{-l}$, $l > 1$, and hence $\mathbf{r}(z)$ are continuous in norm up to the cut along $[0,\infty)$.*

Now it follows from Propositions 12.1 and 12.5 that Proposition 2.3 can be applied to the operator-valued function $\mathbf{r}(z)$ defined by equalities (12.6). Therefore the operator-function $(I+\mathbf{r}(z))^{-1}$ is analytic in the complex plane cut along $[0,\infty)$, and it is continuous up to the cut $[0,\infty)$ except points $\lambda \in [0,\infty)$ where the homogeneous equation

$$\mathbf{f} + \mathbf{r}(\lambda \pm i0)\mathbf{f} = 0 \tag{12.18}$$

has a non-trivial solution. The set $\mathcal{N} = \mathcal{N}_+ \cup \mathcal{N}_-$ of such points $\lambda \in [0,\infty)$ is closed and has Lebesgue measure zero.

Multiplying from the right equations (12.4) by $\langle x^\gamma \rangle^{-l}$ for all $\gamma = (12), (23), (31)$, we get

Proposition 12.6 *The operator-functions $\langle x^\alpha \rangle^{-l} R(z) \langle x^\beta \rangle^{-l}$, $l > 1$, are norm-continuous as the complex parameter z approaches the positive half-axis at the points of $\Lambda = [0,\infty) \setminus \mathcal{N}$. In particular, the operators $\langle x^\alpha \rangle^{-l}$, $l > 1$, are H-smooth on any compact subinterval of Λ. The operators $\langle x^\alpha \rangle^{-l} R(z) \langle x^\beta \rangle^{-l}$, $\alpha \neq \beta$, are compact.*

Combining Propositions 12.2, 12.6 with Proposition 1.18, we obtain

Theorem 12.7 *Let $k \geq 3$ and let condition* (11.15) *on pair potentials V^α be fulfilled for some $\rho > 2$. Suppose that all three operators H^α do not have eigenvalues and zero-energy resonances. Then the wave operators $W^\pm(H, H_0)$ exist and are complete.*

12.3. The multichannel case. Next we consider a three-particle system with the operators H^α having negative spectrum. To simplify notation, we assume that every H^α has exactly one eigenvalue $\lambda^\alpha < 0$. The corresponding normalized eigenfunction $\psi^\alpha(x^\alpha)$ decays exponentially at infinity. In the general case system (12.4) should be further rearranged. Let the operators K_α and

$$J^\alpha : L_2(\mathbf{R}^k) =: \mathfrak{h} \to \mathcal{H} := L_2(X)$$

be defined by equalities (11.10) and (11.20), that is

$$K_\alpha = -(2m_\alpha)^{-1}\Delta_{x_\alpha}, \quad J^\alpha f_\alpha = f_\alpha \otimes \psi^\alpha. \tag{12.19}$$

The operator $\langle x^\alpha \rangle^l J^\alpha$ is bounded for any l. Set $P_\alpha = J^\alpha (J^\alpha)^*$ and $\tilde{R}_\alpha(z) = (I - P_\alpha) R_\alpha(z)$. Clearly, (cf. (12.12))

$$P_\alpha R_\alpha(z) = J^\alpha (K_\alpha + \lambda^\alpha - z)^{-1} (J^\alpha)^*.$$

It follows from the resolvent identity for the pair H_0, H_α that

$$\tilde{R}_\alpha(z) = (I - P_\alpha) R_0(z) - \tilde{R}_\alpha(z) V^\alpha R_0(z). \tag{12.20}$$

Let us split up $\mathcal{R}_\alpha(z) = \langle x^\alpha \rangle^{-l} R(z)$ into two components

$$\left.\begin{aligned} \mathcal{R}_{\alpha 0}(z) &= \langle x^\alpha \rangle^{-l} (I - P_\alpha) R(z), \\ \mathcal{R}_{\alpha 1}(z) &= \langle x_\alpha \rangle^{l} (J^\alpha)^* - \langle x_\alpha \rangle^{l} (J^\alpha)^* \textstyle\sum_{\beta \neq \alpha} V^\beta R(z). \end{aligned}\right\} \tag{12.21}$$

Applying the operators P_α to the resolvent identities (12.3), we see that

$$P_\alpha R(z) = J^\alpha (K_\alpha + \lambda^\alpha - z)^{-1} \langle x_\alpha \rangle^{-l} \mathcal{R}_{\alpha 1}(z)$$

and hence

$$\mathcal{R}_\alpha(z) = \mathcal{R}_{\alpha 0}(z) + \langle x^\alpha \rangle^{-l} J^\alpha (K_\alpha + \lambda^\alpha - z)^{-1} \langle x_\alpha \rangle^{-l} \mathcal{R}_{\alpha 1}(z). \tag{12.22}$$

The second equality (12.21) will be considered as an equation for the operator $\mathcal{R}_{\alpha 1}(z) : \mathcal{H} \to \mathfrak{h}$. Applying the operators $\langle x^\alpha \rangle^{-l}(I - P_\alpha)$ to (12.3), we obtain also equations for components $\mathcal{R}_{\alpha 0}(z) : \mathcal{H} \to \mathcal{H}$. Thus we have a system of 6 equations for 6 operators $\mathcal{R}_{\alpha 0}(z)$ and $\mathcal{R}_{\alpha 1}(z)$:

$$\left.\begin{array}{rcl} \mathcal{R}_{\alpha 0}(z) & = & \langle x^\alpha \rangle^{-l} \tilde{R}_\alpha(z) - \langle x^\alpha \rangle^{-l} \tilde{R}_\alpha(z) \sum_{\beta \neq \alpha} \langle x^\beta \rangle^{-l} B^\beta \mathcal{R}_\beta(z), \\ \mathcal{R}_{\alpha 1}(z) & = & \langle x_\alpha \rangle^{l} (J^\alpha)^* - \langle x_\alpha \rangle^{l} (J^\alpha)^* \sum_{\beta \neq \alpha} \langle x^\beta \rangle^{-l} B^\beta \mathcal{R}_\beta(z), \end{array}\right\} \tag{12.23}$$

where $\mathcal{R}_\alpha(z)$ are given by formula (12.22).

We consider (12.23) again as a vector equation of the form (12.5). The corresponding matrix operator $\mathbf{r}(z)$ acts in three copies of $\mathcal{H}$ and three copies of $\mathfrak{h}$. According to (12.22), (12.23) its components $\mathbf{r}_{\alpha j, \alpha k}(z) = 0$ for all $\alpha = (12), (23), (31)$ and $j, k = 0, 1$. If $\alpha \neq \beta$, then

$$\left.\begin{array}{rcl} \mathbf{r}_{\alpha 0, \beta 0}(z) & = & \langle x^\alpha \rangle^{-l} \tilde{R}_\alpha(z) \langle x^\beta \rangle^{-l} B^\beta, \\ \mathbf{r}_{\alpha 0, \beta 1}(z) & = & \langle x^\alpha \rangle^{-l} \tilde{R}_\alpha(z) V^\beta J^\beta (K_\beta + \lambda^\beta - z)^{-1} \langle x_\beta \rangle^{-l}, \\ \mathbf{r}_{\alpha 1, \beta 0} & = & \langle x_\alpha \rangle^{l} (J^\alpha)^* \langle x^\beta \rangle^{-l} B^\beta, \\ \mathbf{r}_{\alpha 1, \beta 1}(z) & = & \langle x_\alpha \rangle^{l} (J^\alpha)^* V^\beta J^\beta (K_\beta + \lambda^\beta - z)^{-1} \langle x_\beta \rangle^{-l}. \end{array}\right\}$$

Let $f = \{f_{\alpha 0}, f_{\alpha 1}\}$ satisfy the homogeneous equation $f = -\mathbf{r}(z) f$, or

$$\left.\begin{array}{rcl} f_{\alpha 0} & = & -\langle x^\alpha \rangle^{-l} \tilde{R}_\alpha(z) \sum_{\beta \neq \alpha} \langle x^\beta \rangle^{-l} B^\beta f_\beta, \\ f_{\alpha 1} & = & -\langle x_\alpha \rangle^{l} (J^\alpha)^* \sum_{\beta \neq \alpha} \langle x^\beta \rangle^{-l} B^\beta f_\beta, \end{array}\right\} \tag{12.24}$$

where

$$f_\alpha = f_{\alpha 0} + \langle x^\alpha \rangle^{-l} J^\alpha (K_\alpha + \lambda^\alpha - z)^{-1} \langle x_\alpha \rangle^{-l} f_{\alpha 1}.$$

Then f_α satisfy system (12.8). Therefore, by Proposition 12.1, $f_\alpha = 0$ for all α. Finally, system (12.24) implies that $f_{\alpha 0} = 0$ and $f_{\alpha 1} = 0$.

Using identity (12.20), we check, similarly to subsections 12.1 and 12.2, that the operator-function $\mathbf{r}_{\alpha 0, \beta 0}(z)$ takes compact values, it is analytic in the complex plane cut along $[0, \infty)$ and is continuous in norm up to the cut. By inequality

$$\langle x_\alpha \rangle \leq C \langle x^\alpha \rangle \langle x^\beta \rangle, \quad \alpha \neq \beta,$$

the operators $\mathbf{r}_{\alpha 1, \beta 0}$ and $\langle x_\alpha \rangle^{l} (J^\alpha)^* V^\beta J^\beta \langle x_\beta \rangle^{l}$ are bounded. Since the operator

$$\langle x_\beta \rangle^{-l} (K_\beta + \lambda^\beta - z)^{-1} \langle x_\beta \rangle^{-l} \tag{12.25}$$

is compact in $\mathfrak{h}$ and continuous up to the cut $[\lambda^\beta, \infty)$ (see subsection 3.1), the operator $\mathbf{r}_{\alpha 1, \beta 1}(z)$ also has all these properties.

It remains to consider the operator $\mathbf{r}_{\alpha 0,\beta 1}(z)$. Note that the operators $V^\beta J^\beta$ and $(K_\beta + \lambda^\beta - z)^{-1}$ commute. Taking into account identity (12.20), we see that

$$\tilde{R}_\alpha(z) V^\beta J^\beta (K_\beta + \lambda^\beta - z)^{-1} = (I - P_\alpha - \tilde{R}_\alpha(z) V^\alpha) R_0(z) (K_\beta + \lambda^\beta - z)^{-1} V^\beta J^\beta,$$

where

$$R_0(z)(K_\beta + \lambda^\beta - z)^{-1} = ((K_\beta + \lambda^\beta - z)^{-1} - R_0(z))(-(2m^\beta)^{-1}\Delta_{x^\beta} + \lambda^\beta)^{-1}.$$

Further, it follows from equation $H^\beta \psi^\beta = \lambda^\beta \psi^\beta$ that

$$(-(2m^\beta)^{-1}\Delta_{x^\beta} + \lambda^\beta)^{-1} V^\beta J^\beta = -J^\beta.$$

Combining these equations, we get the identity

$$\tilde{R}_\alpha(z) V^\beta J^\beta (K_\beta + \lambda^\beta - z)^{-1} = \tilde{R}_\alpha(z) J^\beta - (I - P_\alpha - \tilde{R}_\alpha(z) V^\alpha) J^\beta (K_\beta + \lambda^\beta - z)^{-1}.$$

Thus using the above results on the operators $\langle x^\alpha \rangle^{-l} \tilde{R}_\alpha(z) \langle x^\beta \rangle^{-l}$ and (12.25), we see that the operator-function $\mathbf{r}_{\alpha 0,\beta 1}(z)$ is compact and continuous up to the cut $[\lambda^\beta, \infty)$.

Let us summarize the results obtained.

Proposition 12.8 *The number -1 is not an eigenvalue of the operator $\mathbf{r}(z)$, and the operator $\mathbf{r}^2(z)$ is compact. The operator-function $\mathbf{r}(z)$ is analytic in the complex plane cut along $[\lambda_0, \infty)$, where $\lambda_0 = \{\min \lambda^\alpha\}$, and it is continuous in norm up to the cut.*

Now it follows from Proposition 2.3 that the operator-function $(I + \mathbf{r}(z))^{-1}$ is also continuous in norm up to the cut except the points λ where equation (12.18) has a non-trivial solution. As before, the set $\mathcal{N}$ of such points is closed and has Lebesgue measure zero.

Let us set

$$\mathsf{G}_{\alpha 0} = \langle x^\alpha \rangle^{-l}(I - P_\alpha), \quad \mathsf{G}_{\alpha 1} = \langle x_\alpha \rangle^{l} (J^\alpha)^* \sum_{\beta \neq \alpha} V^\beta.$$

Similarly to the study of $\mathbf{r}(z)$, we can check that, for all α, γ and $j = 0, 1$, the operators $\langle x_\alpha \rangle^l (J^\alpha)^* \mathsf{G}_{\gamma j}^*$ are bounded and the operators $\langle x^\gamma \rangle^{-l} \tilde{R}_\alpha(z) \mathsf{G}_{\gamma j}^*$ are bounded and norm continuous up to the cut. Multiplying from the right all equations (12.23) by $\mathsf{G}_{\gamma 0}^*$ and $\mathsf{G}_{\gamma 1}^*$ for all $\gamma = (12), (23), (31)$ and collecting the results obtained together, we get

Proposition 12.9 *Let $k \geq 3$ and let condition (11.15) on pair potentials V^α be fulfilled for some $\rho > 2$. Suppose that all three operators H^α do not have zero-energy resonances and have exactly one eigenvalue which is negative. Then, for all α, β and $i, j = 0, 1$, the operator-functions $\mathsf{G}_{\alpha i} R(z) \mathsf{G}_{\beta j}^*$ are norm continuous as z approaches the cut $[\lambda_0, \infty)$ at the points of $\Lambda = [\lambda_0, \infty) \setminus \mathcal{N}$. In particular, the operators $\mathsf{G}_{\alpha i}$ are H-smooth on any compact subinterval of Λ.*

12.4. Wave operators and their completeness. If the operators H^α have negative eigenvalues, then, to fit scattering for the Hamiltonian H into the framework of smooth theory, we need to introduce a new identification

$$\mathbf{J}^0 = I - \sum_\alpha P_\alpha. \tag{12.26}$$

A direct calculation shows that

$$\begin{aligned} \mathbf{V}_0 := H\mathbf{J}^0 - \mathbf{J}^0 H_0 &= \sum_\alpha (I - P_\alpha)V^\alpha - \sum_\alpha \Big(\sum_{\beta\neq\alpha} V^\beta\Big) P_\alpha \\ &= \sum_\alpha \mathsf{G}^*_{\alpha 0} B^\alpha \langle x^\alpha\rangle^{-l} + \sum_\alpha \mathsf{G}^*_{\alpha 1}\langle x_\alpha\rangle^{-l}(J^\alpha)^*. \end{aligned} \tag{12.27}$$

By Proposition 12.2, the operators $\langle x^\alpha\rangle^{-l}$, $l > 1$, and consequently $(J^\alpha)^*$ are H_0-smooth. Thus it follows from Proposition 12.9 that every term in (12.27) can be factored into a product of H_0-smooth and H-smooth operators. A verification of a similar statement for the perturbation

$$\mathbf{V}_\alpha := HJ^\alpha - J^\alpha \mathbf{K}_\alpha = \Big(\sum_{\beta\neq\alpha} V^\beta\Big) J^\alpha, \quad \mathbf{K}_\alpha = K_\alpha + \lambda^\alpha, \tag{12.28}$$

is quite straightforward. Applying again Proposition 1.18, we arrive at

Theorem 12.10 *Under the assumptions of Proposition* 12.9, *the wave operators*

$$W^\pm(H, H_0; I - \sum P_\alpha), \quad W^\pm(H_0, H; I - \sum P_\alpha)$$

and $W^\pm(H, \mathbf{K}_\alpha; J^\alpha)$, $W^\pm(\mathbf{K}_\alpha, H; (J^\alpha)^*)$ *exist.*

It is convenient to reformulate the result obtained in terms of scattering theory in a couple of spaces. Let the space $\hat{\mathcal{H}}$, the operator $\hat{H}$ and the identification $\hat{J}$ be defined by equalities (11.18), (11.19) and (11.20), that is

$$\hat{\mathcal{H}} = \mathcal{H}_0 \oplus \bigoplus_\alpha \mathcal{H}_\alpha, \quad \mathcal{H}_0 = L_2(\mathbf{R}^{2k}), \quad \mathcal{H}_\alpha = L_2(\mathbf{R}^k),$$

$$\hat{H} = H_0 \oplus \bigoplus_\alpha \mathbf{K}_\alpha, \quad \hat{J} = I \oplus \bigoplus_\alpha J^\alpha.$$

We introduce now a new identification by the formula

$$\hat{\mathbf{J}} = \mathbf{J}^0 \oplus \bigoplus_\alpha J^\alpha,$$

where $\mathbf{J}^0$ is operator (12.26). Theorem 12.10 shows that the wave operators

$$W^\pm(H, \hat{H}; \hat{\mathbf{J}}) \quad \text{and} \quad W^\pm(\hat{H}, H; \hat{\mathbf{J}}^*)$$

exist.

Our next goal is to check that both of them are isometric. Note first of all that if $f(x) = f_\alpha(x_\alpha) f^\alpha(x^\alpha)$, then

$$||P_\alpha \exp(-iH_0 t) f|| = |(\exp(i(2m^\alpha)^{-1}\Delta_{x^\alpha} t) f^\alpha, \psi^\alpha)| \; ||f_\alpha||$$

tends to zero as $|t| \to \infty$. This implies that

$$s - \lim_{|t|\to\infty} P_\alpha \exp(-iH_0 t) = 0. \tag{12.29}$$

We remark also that

$$P_\alpha P_\beta \in \mathfrak{S}_\infty, \quad \alpha \neq \beta, \tag{12.30}$$

(actually, these operators belong to the Hilbert-Schmidt class).

Let us now show that the pair $\hat{H}$, $\hat{\mathbf{J}}$ satisfies condition (1.9), that is

$$s - \lim_{|t|\to\infty} (\hat{\mathbf{J}}^* \hat{\mathbf{J}} - I) \exp(-i\hat{H}t) = 0. \tag{12.31}$$

Clearly,

$$(\mathbf{J}^0)^* \mathbf{J}^0 - I = -\sum_\alpha P_\alpha + \sum_{\alpha \neq \beta} P_\alpha P_\beta,$$

so that relation (12.31) on the subspace $\mathcal{H}_0$ is a consequence of (12.29), (12.30). All other components of the operator $\hat{\mathbf{J}}^* \mathbf{J} - I$ reduce to $P_\alpha P_\beta$, $\alpha \neq \beta$, and hence are compact. Thus the operator $W^\pm(H, \hat{H}; \hat{\mathbf{J}})$ is isometric.

Next we note that the operator

$$\hat{\mathbf{J}}\hat{\mathbf{J}}^* - I = (I - \sum_\alpha P_\alpha)^2 + \sum_\alpha J^\alpha (J^\alpha)^* - I = \sum_{\alpha \neq \beta} P_\alpha P_\beta$$

is compact and consequently the pair H, $\hat{\mathbf{J}}^*$ satisfies condition (1.9). This implies isometricity of the operator $W^\pm(\hat{H}, H; \hat{\mathbf{J}}^*)$.

Finally, we remark that, by (12.29),

$$W^\pm(H, H_0; \mathbf{J}^0) = W^\pm(H, H_0) \quad \text{and hence} \quad W^\pm(H, \hat{H}; \hat{\mathbf{J}}) = W^\pm(H, \hat{H}, \hat{J}).$$

This concludes the proof of asymptotic completeness.

Theorem 12.11 *Under the assumptions of Proposition* 12.9, *the wave operators* $W^\pm(H, \hat{H}, \hat{J})$ *exist, are isometric and complete.*

Of course, this result is equivalent to the existence of the wave operators

$$W^\pm(H, H_0), \quad W^\pm(H, K_\alpha; J^\alpha),$$

their isometricity, orthogonality of their ranges and the equality

$$\operatorname{Ran} W^\pm(H, H_0) \oplus \bigoplus_\alpha \operatorname{Ran} W^\pm(H, K_\alpha; J^\alpha) = \mathcal{H}^{(ac)}.$$

12.5. Historical remarks. Theorem 12.11 was proven by L. D. Faddeev [48] under somewhat stronger assumptions on pair potentials. Actually, he studied the system of resolvent equations in appropriate Banach spaces and did not use the passage from (12.3) to (12.4). The latter form appeared first in the paper [122] by R. Newton. Under the assumptions of Proposition 12.9 the asymptotic completeness was proven by J. Ginibre, M. Moulin [59] and L. Thomas [149]. They introduced the Hilbert space technique in this context and made essential use of Proposition 12.4 obtained in [76]. Identification (12.26), which allowed us to fit three particle scattering into the standard framework of smooth scattering theory in a couple of spaces, appeared in [155].

According to Proposition 12.9 the singular continuous spectrum of the operator H is contained in the set $\mathcal{N}$ where homogeneous equation (12.18) has not-trivial solutions. However, as shown in [156] under the assumption $\rho > 11/4$, every $\lambda \in \mathcal{N}$ is an eigenvalue of H and hence its singular continuous spectrum is empty.

We emphasize that the proof above requires that all masses of particles be finite. If, for example $m_1 = \infty$, then the left-hand side of (12.16) does not depend on t and the conclusion of Proposition 12.4 about the operator-function $\langle x^{12}\rangle^{-l} R_0(z) \langle x^{31}\rangle^{-l}$ fails.

A generalization of this construction to an arbitrary number of particles meets with numerous problems (see, e.g., [62], [139]). Let us discuss only the case where all subhamiltonians H^a do not have eigenvalues. If $N > 3$, then system (12.4) is no longer Fredholm and has to be further rearranged. A necessary modification was found by O. A. Yakubovsky [175]. The modified system can be written in form (12.5) with a compact operator $\mathbf{r}(z)$ for $\operatorname{Im} z \neq 0$. Of course this system is getting more and more complicated as N increases. However, the main difficulty consists in passing to the limit $\operatorname{Im} z \to 0$. For example, one encounters the terms like $V^{(12)} R_0(z) V^{(34)}$ which are not norm-continuous in z although they remain uniformly bounded. This impedes a proof of compactness of the operators $\mathbf{r}(\lambda \pm i0)$. Another problem is that Faddeev's approach to three-particle scattering requires rather a complete information about two-particle Hamiltonians. Similarly, to treat N-particle case, it is not sufficient to use the asymptotic completeness of all subsystems. Unfortunately an additional information on subhamiltonians (for example, on the structure of their resolvents at threshold energies) is also needed. This looks like a complicated and burdensome problem.

13. ASYMPTOTIC COMPLETENESS. A SKETCH OF PROOF

Our aim in this section is to prove Theorem 11.1. We follow here the paper [166].

The approach of this section to N-particle scattering is similar, from analytical point of view, to that of Section 4 for two-particle Hamiltonians with long-range potentials. In both cases we rely on the limiting absorption principle. In the two-particle long-range case the Mourre commutator method is the simplest way, and in the multiparticle case it is the only way, to obtain this analytical result. In both cases the limiting absorption principle is not sufficient for construction of scattering theory, and an additional analytical information is provided by radiation estimates (Theorem 4.3, in the two-particle case).

13.1. The limiting absorption principle and radiation estimates. We consider here the Schrödinger operator $H = -\Delta + V$ with potential (11.6). Recall that V^a is defined by (11.7), (11.9) and the operator $H^a = -\Delta_{x^a} + V^a$ acts in the space $L_2(X^a)$. Eigenvalues $\lambda^{\mathbf{a}}$, $\mathbf{a} = \{a, n\}$, of the operators H^a for all $X^a \subset \mathcal{X}$ are called thresholds for the Hamiltonian H. The set of all thresholds is denoted Υ_0. As before, $\langle x \rangle$ is the operator of multiplication by $(|x|^2+1)^{1/2}$ and $\mathbf{A}$ is the generator of dilations (4.10). The limiting absorption principle as well as many other intermediary results is valid both for short- and long-range "pair" potentials V^α. The following basic result was established in [120], [130] (see also [54], [160] and the books [34], [5]).

Theorem 13.1 *Let each pair potential $V^\alpha = V_S^\alpha + V_L^\alpha$ where V_S^α satisfies assumption* (11.15) *for some $\rho = \rho_0 > 1$ and*

$$|V_L^\alpha(x^\alpha)| + (1+|x^\alpha|)\,|\nabla V_L^\alpha(x^\alpha)| \le C(1+|x^\alpha|)^{-\rho_1}, \quad \rho_1 > 0,$$

respectively. Then the set of thresholds Υ_0 is closed and countable. Eigenvalues of H may accumulate at Υ_0 only, so the "exceptional" set $\Upsilon = \Upsilon_0 \cup \sigma_H^{(p)}$ is also closed and countable. Furthermore, for every $\lambda \in \mathbf{R}\setminus\Upsilon$ there exists a small interval $\Lambda_\lambda \ni \lambda$ such that the Mourre estimate (4.9) *for the commutator $[H, \mathbf{A}]$ holds. Finally, for any compact interval Λ such that $\Lambda \cap \Upsilon = \emptyset$ and any $l > 1/2$, the operator $\langle x \rangle^{-l} E(\Lambda)$ is H-smooth (the limiting absorption principle). In particular, the operator H does not have singular continuous spectrum.*

We emphasize that the proof of the Mourre estimate in the multiparticle case is considerably more complicated than that given in subsection 4.2. In particular, one should avoid eigenvalues of all operators H^a, whereas in the two-particle case Υ_0 consists of the point 0 only. On the other hand, as was already mentioned, the limiting absorption principle is derived from the Mourre estimate by essentially abstract arguments.

Recall (see subsection 2.1) that for two-particle systems with short-range potentials asymptotic completeness is an immediate consequence of the limiting absorption principle. For N-particle ($N \geq 3$) systems, we need also radiation estimates which, in contrast to the two-particle case (Theorem 4.3), look differently in different regions of the configuration space.

Let $\nabla_a = \nabla_{x_a}$ be the gradient in the variable x_a and let $\nabla_a^\perp$,

$$(\nabla_a^\perp u)(x) = (\nabla_a u)(x) - |x_a|^{-2}\langle(\nabla_a u)(x), x_a\rangle x_a,$$

be its orthogonal projection in X_a on the subspace orthogonal to the vector x_a. We introduce a special notation Z_a for X_a with all $X_b \subset X_a$, $X_b \neq X_a$, removed from it, that is

$$Z_a = X_a \setminus \bigcup_{X_b \subset X_a, X_b \neq X_a} X_b.$$

Clearly, $x \in Z_a$ if and only if X_a is the smallest subspace from the set $\mathcal{X}'$ containing the point x. It is easy to see that $Z_a \cap Z_b \subset \{0\}$ if $a \neq b$ and

$$\bigcup_a Z_a = \mathbf{R}^d$$

(we recall that a labels all subspaces X_a except $X_a = \{0\}$). For example, for the Hamiltonian (11.11)

$$Z_0 = \mathbf{R}^3 \setminus \bigcup_{\alpha=1}^{3} X_\alpha, \quad Z_\alpha = X_\alpha \setminus \bigcup_{\beta \neq \alpha} X_{\alpha\beta}, \quad Z_{\alpha\beta} = X_{\alpha\beta}.$$

Let us introduce also a conical neighbourhood of a point $\mathbf{x} \in \mathbb{S}^{d-1}$:

$$\mathsf{C}(\mathbf{x};\varepsilon) = \{x : \langle x, \mathbf{x}\rangle > (1-\varepsilon)|x|\}.$$

Now we can formulate our main analytical result.

Theorem 13.2 *Suppose that the assumptions of Theorem* 13.1 *hold. If* $\mathbf{x}_a \in Z_a \cap \mathbb{S}^{d-1}$, *then for sufficiently small* $\varepsilon > 0$ *the operator*

$$G(\mathbf{x}_a, \varepsilon) = \chi(\mathsf{C}(\mathbf{x}_a;\varepsilon))\langle x\rangle^{-1/2}\nabla_a^\perp \tag{13.1}$$

is H-smooth on Λ.

As the two-particle case, we refer to the estimates of Theorem 13.2 as radiation estimates. By virtue of inequality

$$|(\nabla_b^\perp u)(x)| \leq |(\nabla_a^\perp u)(x)| \quad \text{if} \quad X_b \subset X_a, \tag{13.2}$$

this estimate is stronger for $\mathbf{x}_a \in Z_a$ than for $\mathbf{x}_b \in Z_b$ if $X_b \subset X_a$ and $X_b \neq X_a$. In particular, it is the strongest for $\mathbf{x}_0 \in Z_0$, that is in the "free" region, where all potentials V^α are vanishing. In this region functions $(U(t)f)(x)$ satisfy for large t and x the same estimate as in the two-particle case. On the contrary, $\nabla_a^\perp = 0$

if $\dim X_a = 1$, so Theorem 13.2 is trivial in this case. According to conjecture (11.21), for arbitrary a and $\mathbf{x}_a \in Z_a$, in a small cone $\mathsf{C}(\mathbf{x}_a;\varepsilon)$ evolution of the system is asymptotically free in the variable x_a. Therefore one can expect that for every a the operator $\chi(\mathsf{C}(\mathbf{x}_a;\varepsilon))\nabla_a^\perp$ is "improving". In particular, for the N-particle Hamiltonian $\mathbf{H}$ in the region of the configuration space, where clusters $C_1,\dots,C_p$ are far from each other compared to distances between particles inside each cluster, this means that the motion is asymptotically free in center of mass variables defined by (11.14).

Theorem 13.2 can also be reformulated in a "global" (with respect to directions of x) form although we do not need it for the proof of Theorem 11.1. Let $\mathbf{Y}_a$ be a closed cone in $\mathbb{R}^d$ such that $\mathbf{Y}_a \cap X_b = \{0\}$ if $X_a \not\subset X_b$; in particular, $\mathbf{Y}_0 \cap X_\alpha = \{0\}$ for all X_α. Let $\chi(\mathbf{Y}_a)$ denote the characteristic function of $\mathbf{Y}_a$. Theorem 13.2 is equivalent to

Theorem 13.2 bis *Suppose that the assumptions of Theorem* 13.1 *hold. Then for any a, the operator $G_a = \chi(\mathbf{Y}_a)\langle x\rangle^{-1/2}\nabla_a^\perp$ is H-smooth on Λ.*

Theorem 13.2 bis obviously implies Theorem 13.2. To see the converse, one uses estimate (13.2) and standard covering arguments.

Let us illustrate the definitions above on the example of Hamiltonian (11.11). First, we have to exclude the subspaces $X_{\alpha\beta}$ (because $\dim X_{\alpha\beta} = 1$). If $\alpha = 1,2,3$, then the operator $\chi(\mathbf{Y}_\alpha)\langle x\rangle^{-1/2}\nabla_\alpha^\perp$ is H-smooth for any cone $\mathbf{Y}_\alpha$ such that $\mathbf{Y}_\alpha \cap X_\beta = \{0\}$ for $\beta \neq \alpha$. The operator $\chi(\mathbf{Y}_0)\langle x\rangle^{-1/2}\nabla^\perp$ is H-smooth if $\mathbf{Y}_0 \cap X_\alpha = \{0\}$ for all α. On the picture below, $\mathbf{Y}_0$ is doubly shadowed, $\mathbf{Y}_1 \setminus \mathbf{Y}_0$ is simply shadowed.

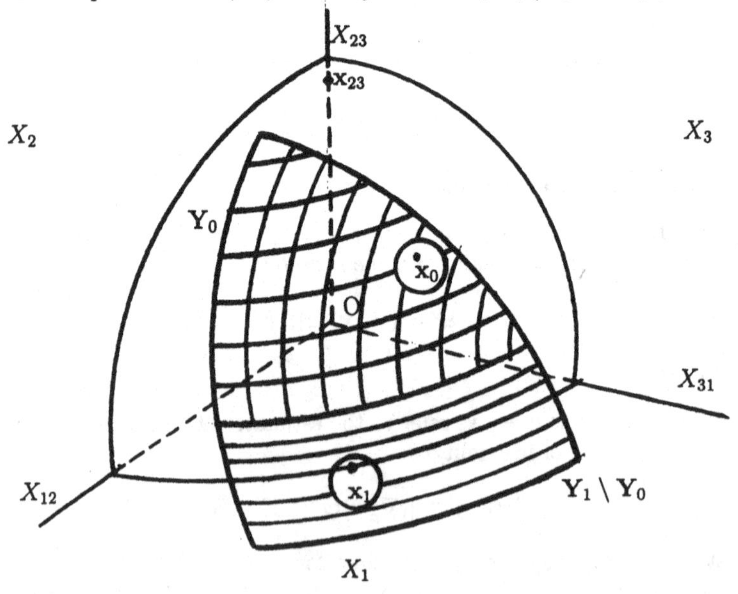

Figure 13.1. The points $\mathbf{x}_a$ and the cones $\mathbf{Y}_a$ for Hamiltonian (11.11).

Similarly to subsection 4.2, for the proof of Theorem 13.2 we proceed from Proposition 1.19. Let us seek M as a first-order differential operator (4.14) determined by the function m which we call "generating". Of course only values of $m(x)$ (and of other auxiliary functions) for large $|x|$ are essential. So we always impose conditions for, say, $|x| \geq 1$ allowing m to be arbitrary (we only require that $m \in C^\infty(\mathbb{R}^d)$) inside the ball $|x| \leq 1$. In the N-particle case we cannot take $m(x) = |x|$ since then estimate (4.19) is violated. This is particularly clear from formula (4.17) for $[V_L^\alpha, M]$. This formula shows that, in order that (4.19) be satisfied, $m(x)$ should not depend on the variables x^α in some neighbourhoods of X_α where $V^\alpha(x^\alpha)$ do not tend to zero.

More precisely, we suppose that m satisfies the following properties:

1^0 $m(x)$ is a real C^∞-function, which is homogeneous of degree 1 and $m(x) \geq 1$.

2^0 For any a, the function $m(x)$ does not depend on x^a, i.e. $m(x) = m(x_a)$, in some conical neighbourhood of X_a.

3^0 $m(x)$ is (locally) convex function, i.e.

$$\sum_{j,k} m^{(jk)}(x)\xi_j\bar{\xi}_k \geq 0, \quad \forall \xi \in \mathbb{C}^d, \quad m^{(j,k)}(x) = \partial^2 m(x)/\partial x_j \partial x_k.$$

4^0 For any point $\mathbf{x}_a \in Z_a \cap \mathbb{S}^{d-1}$ and sufficiently small $\varepsilon > 0$, one can choose $m = m_{\mathbf{x}_a}$ in such a way that

$$m(x) = \mu_a |x_a|, \quad \mu_a \geq 1, \quad x \in \mathsf{C}(\mathbf{x}_a; \varepsilon). \tag{13.3}$$

The concrete construction of a function $m(x)$ satisfying all these properties is given in [166]. Here we remark only that functions

$$m(x, \boldsymbol{\varepsilon}) = \max_a \{(1 + \varepsilon_a)|x_a|\}, \quad \boldsymbol{\varepsilon} = \{\varepsilon_a\},$$

where $\varepsilon_a \in (\epsilon^{d_a}, 2\epsilon^{d_a})$ and ϵ is sufficiently small, satisfy all necessary properties except smoothness. Indeed, 1^0 is obvious; 2^0 is satisfied because, in a small neighbourhood of X_a,

$$(1 + \epsilon^{d_a})|x_a| \geq (1 + 2\epsilon^{d_b})|x_b| \quad \text{for} \quad X_a \subset X_b,$$

if ϵ is small enough; $m(x, \boldsymbol{\varepsilon})$ is convex as the maximum of convex functions. Further, for any $\mathbf{x}_a \in Z_a \cap \mathbb{S}^{d-1}$ we can chose ϵ so small that $m(x, \boldsymbol{\varepsilon}) = \mu_a |x_a|$ in a small neighbourhood of $\mathbf{x}_a$. Actually, $m(x, \boldsymbol{\varepsilon}) = \mu_a |x_a|$ in conical neighbourhoods Y_a of X_a (for all a) with some conical neighbourhoods of all $X_b \subset X_a$, $X_b \neq X_a$, removed from it. Finally, a C^∞-function with the same properties can be defined by the formula

$$m(x) = \int m(x, \boldsymbol{\varepsilon}) \prod_a \varphi_a(\varepsilon_a) d\varepsilon_a \tag{13.4}$$

where $\varphi_a \geq 0$ are suitable smooth functions.

For example, in the case (11.11) $m(x) = \mu_0|x|$ for $x \in Y_0$ (this is always true in the free region), $m(x) = \mu_\alpha|x_\alpha|$ for $x \in Y_\alpha$ if $a = 1, 2, 3$ and $m(x) = \mu_a|x_a|$ for $x \in Y_a$ if $a = 12, 23, 31$. Thus the level surface $m(x) = const$ is a part of the sphere for $x \in Y_0$, it is a part of a cylinder for $x \in Y_\alpha$, $\alpha = 1, 2, 3$, and it is a part of a plane for $x \in Y_a$, $a = 12, 23, 31$; see Figure 13.2.

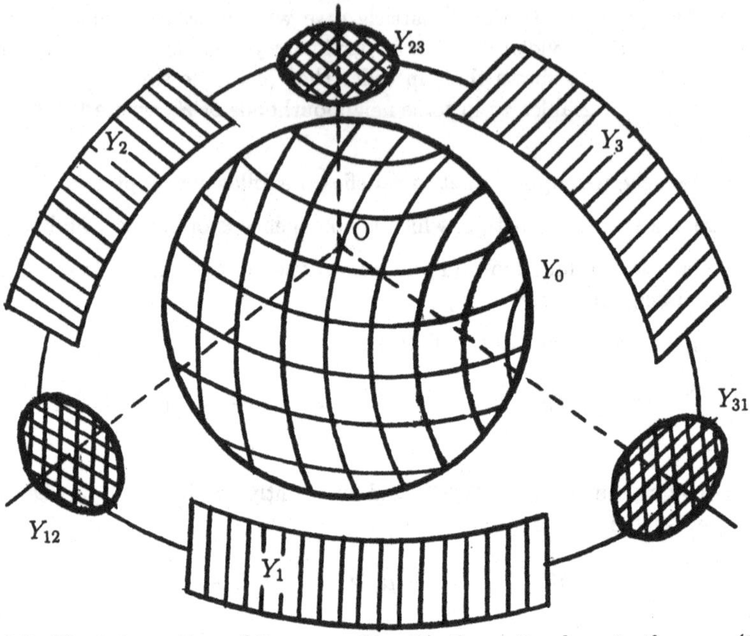

Figure 13.2. The intersection of the cones Y_a with the unit sphere in the case (11.11).

To estimate the commutator $[H_0, M]$, we take into account that the matrix $\{m^{(j,k)}(x)\}$ is nonnegative for $|x| \geq 1$. Therefore by virtue of (4.15)

$$i([H_0, M]u, u) \geq 4\sum_{j,k} \int_C m^{(j,k)} u_j \bar{u}_k dx - c\int_{|x|<1} |\nabla u|^2 dx - c\|\langle x\rangle^{-3/2}u\|^2, \quad (13.5)$$

where $u_j = \partial u/\partial x_j$ and C is any region lying outside of the unit ball. Using identity (4.16) in the variable x_a, we see that

$$\sum_{j,k} m^{(j,k)} u_j \bar{u}_k = \mu_a|x_a|^{-1}|\nabla_a^\perp u|^2 \quad (13.6)$$

if $m(x)$ is defined by (13.3).

Estimate (4.19) follows from

Lemma 13.3 *Let m be a C^∞-function which is homogeneous of degree* 1 *and such that $m(x) = m(x_\alpha)$ in some conical neighbourhood Y_α of X_α. Then*

$$|([V^\alpha, M]u, u)| \leq C\|\langle x\rangle^{-l}(H_0 + I)^{1/2}u\|^2,$$

where l is defined by (4.20).

Indeed, for the long-range part V_L^α we use formula (4.17). If $x \in Y_\alpha$, then $\langle \nabla m(x_\alpha), \nabla V_L^\alpha(x^\alpha)\rangle = 0$ since $\nabla m(x_\alpha)$ belongs to the subspace X_α and $\nabla V_L^\alpha(x^\alpha) = 0$ is orthogonal to it. If $|x| \to \infty$ off Y_α, then $\nabla V_L^\alpha(x^\alpha) = O(|x|^{-1-\rho_1})$. To treat the short-range part V_S^α, we introduce a homogeneous function ζ of degree 0 such that $\operatorname{supp}\zeta \subset Y_\alpha$ and $\zeta(x) = 1$ in some conical neighbourhood $Y'_\alpha \subset Y_\alpha$ of X_α. Since $V_S^\alpha(x^\alpha) = O(|x|^{-\rho_0})$ off any conical neighbourhood of X_α, similarly to the two-particle case (cf. (4.18)), we have that

$$|((1-\zeta)V_S^\alpha M u, u)| + |((1-\zeta)M V_S^\alpha u, u)| \le C\|\langle x\rangle^{-\rho_0/2}(H_0+I)^{1/2}u\|^2.$$

So it remains to use that the operator

$$\zeta M = -2i\zeta\langle \nabla m, \nabla_{x_\alpha}\rangle - i\zeta\Delta m,$$

commutes with V_S^α.

Thus taking into account property 4^0 of m and combining (13.5), (13.6) and (4.19), we arrive, for any $\mathbf{x}_a \in Z_a \cap \mathbb{S}^{d-1}$ and ε small enough, at the estimate

$$i([H,M]u,u) \ge 4\mu_a\|G(\mathbf{x}_a,\varepsilon)u\|^2 - c^2\|\langle x\rangle^{-l}(H_0+I)^{1/2}u\|^2, \quad l > 1/2. \qquad (13.7)$$

This is exactly estimate (1.32) for $2G(\mathbf{x}_a,\varepsilon)$, $K = c\langle x\rangle^{-l}(H_0+I)^{1/2}$, and H-bounded operator (4.14). Now let us take into account that, by Theorem 13.1, the operator $\langle x\rangle^{-l}(H_0+I)^{1/2}E(\Lambda)$ is H-smooth. Thus Proposition 1.19 yields that the operator $GE(\Lambda)$ is also H-smooth. This proves Theorem 13.2.

13.2. Auxiliary wave operators. For the proof of asymptotic completeness we first consider auxiliary wave operators

$$W^\pm(H, H_a; M_a E_a(\Lambda)), \quad W^\pm(H_a, H; M_a E(\Lambda)), \qquad (13.8)$$

where the "identifications" M_a are again first-order differential operators (4.15) with suitably chosen "generating" functions $m_a(x)$ and $E_a(\Lambda) = E_{H_a}(\Lambda)$. Recall that a labels all Hamiltonians (11.10) corresponding to $X^a \ne X$. We need the following properties of these functions (as always for $|x| \ge 1$):

1^0 $m_a(x)$ is a real C^∞-function, which is homogeneous of degree 1.

2^0 For any b, the function $m_a(x)$ does not depend on x^b, i.e. $m_a(x) = m_a(x_b)$, in some conical neighbourhood of X_b.

3^0 Let $X_a \not\subset X_b$. Then $m_a(x) = 0$ in some conical neighbourhood of X_b.

4^0 Set

$$m(x) = \sum_a m_a(x). \qquad (13.9)$$

Then $m(x) \ge 1$.

Given a function $m(x)$ satisfying conditions 1^0 and 2^0 of the previous subsection, functions $m_a(x)$ can be constructed by the formula $m_a(x) = \eta_a(x)m(x)$ where $\eta_a(x)$ is a homogeneous (for $|x| \ge 1$) function of degree 0 satisfying properties 2^0 and 3^0

above. Then property 4^0 is fulfilled if a family η_a is a partition of unity on the unit sphere.

Note that $m_0(x) = 0$ in neighbourhoods of all X_b, $X_b \neq X$. In the case (11.11), $m_1(x) = m_1(x_1)$ in a neighbourhood of X_1 and $m_1(x) = 0$ in neighbourhoods of X_2 and X_3; $m_{23}(x) = m_{23}(x_{23})$, $m_{23}(x) = m_{23}(x_2)$, $m_{23}(x) = m_{23}(x_3)$ in some neighbourhoods of X_{23}, X_2, X_3, respectively, and $m_{23}(x) = 0$ in a neighbourhood of X_1. On the picture below, we mark supports of m_1 and m_{23}. In the doubly shadowed parts m_1 depends on x_1 only and m_{23} on x_{23} only. In the simply shadowed parts m_{23} depends either on x_2 (left piece) or x_3 (right piece) only.

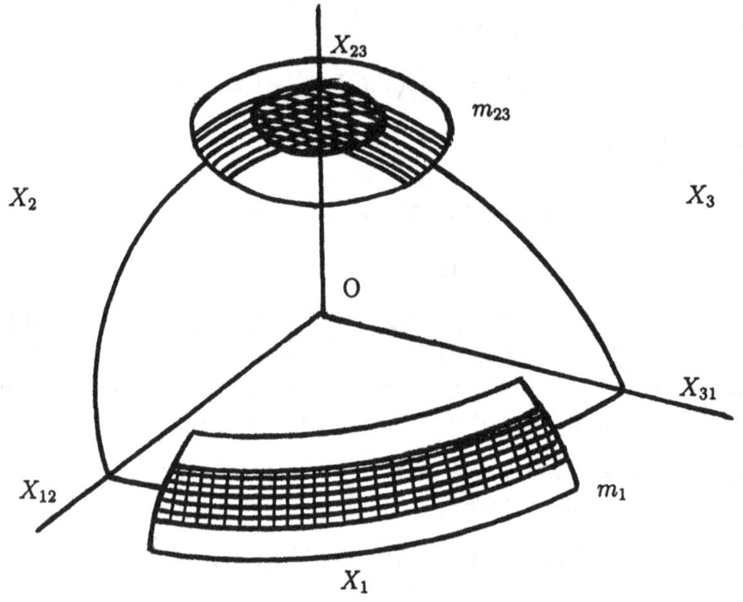

Figure 13.3. Functions m_1 and m_{23} for the case (11.11).

Functions $m_a(x)$ can be constructed similarly to $m(x)$. To that end one first defines functions

$$m_a(x, \varepsilon) = (1 + \varepsilon_a)|x_a|\theta\Big((1 + \varepsilon_a)|x_a| - \max_{b \neq a}\{(1 + \varepsilon_b)|x_b|\}\Big)$$

where $\theta(s) = 1$ for $s \geq 0$ and $\theta(s) = 0$ for $s < 0$. They satisfy all properties formulated above except smoothness. Then $m_a(x)$ is obtained from $m_a(x, \varepsilon)$ by regularization (cf. (13.4)).

Theorem 13.4 *Let a function $m_a(x)$ satisfy properties $1^0 - 3^0$. Then under assumption* (11.15) *the wave operators* (13.8) *exist.*

To prove this theorem, we verify that the triple $H_a = -\Delta + V^a, H, M_a$ satisfies on Λ the conditions of Proposition 1.18. Actually, according to (11.9), the "effective perturbation"

$$T_a := HM_a - M_aH_a = [H_0, M_a] + \sum_{X_a \subset X_\alpha} [V^\alpha, M_a] + \sum_{X_a \not\subset X_\alpha} V^\alpha M_a. \qquad (13.10)$$

To treat the second and third terms we use only H- and H_a-smoothness of the operator $\langle x\rangle^{-l}$, $l > 1/2$. Indeed, the estimate

$$([V^\alpha, M_a]u, u) \leq C||\langle x\rangle^{-l}(H_0 + I)^{1/2}u||^2, \quad 2l = \rho,$$

follows from property 2^0 of m_a and Lemma 13.3. The same estimate for $(V^\alpha M_a u, u)$, $X_a \not\subset X_\alpha$, is a consequence of property 3^0 of m_a and of the bound $V^\alpha(x^\alpha) = O(|x|^{-\rho})$ off any conical neighbourhood of X_α. We emphasize that the short-range assumption is used for the estimate of this term only.

The commutator $i[H_0, M_a]$ is defined by (4.15) with m replaced by m_a. The function $(\Delta^2 m_a)(x) = O(|x|^{-3})$ can again be taken into account by Theorem 13.1. On the contrary, in order to estimate the operator

$$L_a = \sum_{j,k} D_j m_a^{(jk)} D_k, \quad m_a^{(jk)} = \partial^2 m_a / \partial x_j \partial x_k,$$

we have to use Theorem 13.2. Let $\nu_a^{(n)}(x)$ and $p_a^{(n)}(x)$ be eigenvalues and normalized eigenvectors of the symmetric matrix $\mathbf{M}_a(x) = m_a^{(jk)}(x)\}$. Clearly, $\nu_a^{(n)}(x)$ are homogeneous functions of order -1 and $p_a^{(n)}(x)$ – of order 0. Diagonalizing $\mathbf{M}_a(x)$ we find that $L_a = (K_a^{(2)})^* K_a^{(1)}$, where

$$(K_a^{(j)}u)(x) = \sum_{n=1}^{d} s_a^{(n,j)}(x) \langle \nabla u(x), p_a^{(n)}(x)\rangle \, p_a^{(n)}(x), \quad j = 1, 2,$$

and $s_a^{(n,1)} = |\nu_a^{(n)}|^{1/2}$, $s_a^{(n,2)} = |\nu_a^{(n)}|^{1/2} \operatorname{sgn} \nu_a^{(n)}$. To prove H- and H_a-smoothness of the operators $K_a^{(j)}$ we need the following elementary observation based on the Euler theorem.

Lemma 13.5 *Suppose that $m(x) = m(x_b)$ is a smooth homogeneous function of degree 1 in some cone C. Let $\nu^{(n)}(x)$ and $\mathrm{p}^{(n)}(x)$ be eigenvalues and eigenvectors of the symmetric matrix $\mathbf{M}(x) = \{m^{(jk)}(x)\}$. Then vectors $\mathrm{p}^{(n)}(x)$, $x \in \mathsf{C}$, corresponding to $\nu^{(n)}(x) \neq 0$, belong to X_b and are orthogonal to x_b.*

If $\mathbf{x}_b \in Z_b \cap \mathbb{S}^{d-1}$ and ε is small enough, then, by property 2^0, $m_a(x) = m_a(x_b)$ for $x \in \mathsf{C}(\mathbf{x}_b, \varepsilon)$. Hence Lemma 13.5 implies that

$$\langle \nabla u(x), p_a^{(n)}(x)\rangle = \langle \nabla_b^\perp u(x), p_a^{(n)}(x)\rangle, \quad \nu_a^{(n)}(x) \neq 0, \, x \in \mathsf{C}(\mathbf{x}_b, \varepsilon).$$

It follows that

$$|(K_a^{(j)}u)(x)| \leq C\langle x\rangle^{-1/2} |(\nabla_b^\perp u)(x)|, \quad j = 1, 2, \quad x \in \mathsf{C}(\mathbf{x}_b, \varepsilon).$$

Thus H- and H_a-smoothness of the operators $\chi(\mathsf{C}(\mathbf{x}_b, \varepsilon))K_a^{(j)}$ is ensured by Theorem 13.2. Choosing a finite covering of the unit sphere by open sets $\mathsf{C}(\mathbf{x}_b, \varepsilon) \cap \mathbb{S}^{d-1}$, we obtain H- and H_a-smoothness of the operators $K_a^{(j)}$. Putting all things together we arrive at Theorem 13.4.

13.3. A new observable. Our goal now is to deduce Theorem 11.1 from Theorem 13.4. Theorem 13.4 ensures that for any $g \in E(\Lambda)\mathcal{H}$ and every a

$$M_a \exp(-iHt)g = \exp(-iH_a t)g_a^{\pm} + o(1), \quad t \to \pm\infty, \tag{13.11}$$

where

$$g_a^{\pm} = W^{\pm}(H_a, H; M_a E(\Lambda))g.$$

It follows from (13.11) that

$$M \exp(-iHt)g = \sum_a \exp(-iH_a t)g_a^{\pm} + o(1), \quad t \to \pm\infty, \tag{13.12}$$

where

$$\sum_a M_a = M$$

and the sums are taken over all a. If M were the identity operator I (this situation was discussed in [37]), then (13.12) would have easily implied (see below) the asymptotic completeness. However, M is the first-order differential operator, and of course $M \neq I$.

To overcome this difficulty, we introduce the observable

$$M^{\pm}(\Lambda) := W^{\pm}(H, H; ME(\Lambda)). \tag{13.13}$$

Its basic properties are formulated in the following

Proposition 13.6 *Suppose that functions V^{α} satisfy assumptions of Theorem* 13.1. *Let M be operator* (4.14) *with a function m obeying conditions* 1^0 *and* 2^0 *of subsection* 13.1. *Then the wave operators* (13.13) *exist, are self-adjoint and commute with H. Furthermore, their ranges*

$$\operatorname{Ran} M^{\pm}(\Lambda) = E(\Lambda)\mathcal{H}. \tag{13.14}$$

The existence of the wave operators (13.13) can be verified quite similarly to Theorem 13.4. The intertwining property of wave operators means that $M^{\pm}(\Lambda)$ commutes with H. We shall show that $\pm M^{\pm}(\Lambda)$ is positively definite on the subspace $E(\Lambda)\mathcal{H}$. Note the identity

$$d(mf_t, h_t)/dt = i([H, m]f_t, h_t) = i([H_0, m]f_t, h_t) = (Mf_t, h_t), \tag{13.15}$$

where $f_t = \exp(-iHt)f$ and $h_t = \exp(-iHt)h$. Here $h \in \mathcal{H}$ is arbitrary and $f = \psi(H)g$ for some $\psi \in C_0^{\infty}(\Lambda)$ and $g \in \mathcal{D}(\langle x \rangle)$. The set of such elements f is dense in $E(\Lambda)\mathcal{H}$, and mf_t are well-defined. Integrating (13.15) and taking into account the existence of wave operators (13.13) we find that

$$M^{\pm}(\Lambda)f = t^{-1} \exp(iHt) m \exp(-iHt)f + o(1), \quad t \to \pm\infty, \tag{13.16}$$

and, in particular,

$$\|M^{\pm}(\Lambda)f\| = |t|^{-1}\|mf_t\| + o(1). \tag{13.17}$$

Since $m(x) \geq 0$, (13.16) ensures that $\pm M^{\pm}(\Lambda) \geq 0$. To prove that $\pm M^{\pm}(\Lambda) \geq c > 0$, we use Theorem 13.1. By virtue of the identity $i[H, \langle x\rangle^2] = 2\mathbf{A}$, it follows from (4.9) that

$$2^{-1}d^2(\langle x\rangle^2 f_t, f_t)/dt^2 = d(\mathbf{A}f_t, f_t)/dt = (i[H,\mathbf{A}]f_t, f_t) \geq c_0 \, \|f\|^2.$$

Integrating twice this inequality we find that for sufficiently large $|t|$

$$\|\langle x\rangle f_t\| \geq c\,|t|\;\|f\|. \tag{13.18}$$

Comparing (13.17) with (13.18) and considering that $m(x) \geq |x|$ for $|x| \geq 1$, we obtain the inequality

$$\|M^{\pm}(\Lambda)f\| \geq c\,\|f\|, \quad c = c(\Lambda) > 0.$$

Thus $\pm M^{\pm}(\Lambda)$ is positively definite on $E(\Lambda)\mathcal{H}$. In particular, (13.14) holds.

Now we are able to check that for any $f \in E(\Lambda)\mathcal{H}$ and some $f_a^{\pm}$

$$\exp(-iHt)f = \sum_a \exp(-iH_a t)f_a^{\pm} + o(1), \quad t \to \pm\infty. \tag{13.19}$$

Clearly, function (13.9) satisfies the conditions of Proposition 13.6. Therefore $f = M^{\pm}(\Lambda)g$, for some $g = g^{\pm} \in E(\Lambda)\mathcal{H}$, so the asymptotic relation

$$\exp(-iHt)f = M\exp(-iHt)g + o(1), \quad t \to \pm\infty, \tag{13.20}$$

holds. Comparing (13.12) with (13.20), we arrive at relation (13.19) with $f_a^{\pm} = g_a^{\pm}$.

Now it easy to prove (11.21) by inductive arguments. Supposing that Corollary 11.3 is true for all operators H^a and separating the variables according to (11.10), we deduce from (13.19) relation (11.21). Note that by its proof we have used only the existence of the second set of wave operators (13.8).

It remains to prove the existence of wave operators (11.16). Let us define on X_a the function

$$\underline{m}_a(x_a) = |x_a| m_a(x_a|x_a|^{-1}).$$

This function concides with the restriction of $m_a(x)$ to the subspace X_a if $|x_a| \geq 1$, and it is extended by homogeneity (of degree 1) to all $x_a \neq 0$. Using asymptotic relations (3.3), (3.4) for the free evolution $\exp(-iH_0 t)f$, it is easy to check that

$$(\mathcal{F}_0 W^{\pm}(H_0, H_0; M_0 E_0(\Lambda))f)(\xi) = \pm 2\underline{m}_0(\pm\xi)\chi_\Lambda(|\xi|^2)\hat{f}(\xi) \tag{13.21}$$

($\mathcal{F}_0$ is the Fourier transform) for an arbitrary smooth homogeneous function m_0 of degree 1 and any compact $\Lambda \subset (0,\infty)$. This relation can be generalized to all a. Denote by $\mathcal{F}_a$ the Fourier transform in the variable x_a.

Lemma 13.7 *Let $m_a(x)$ be an arbitrary smooth homogeneous function of degree 1 and $m_a(x) = m_a(x_a)$ in some conical neighbourhood of X_a for $|x| \geq 1$. Define the operator $\Omega^{\pm}(m_a) : L_2(X_a) \to L_2(X_a)$ by the equality*

$$(\mathcal{F}_a\Omega^{\pm}(m_a)f_a)(\xi_a) = \pm 2\underline{m}_a(\pm\xi_a)(\mathcal{F}_a f_a)(\xi_a). \tag{13.22}$$

Then for any $\mathbf{a} = \{a, n\}$ and any compact $\Lambda \subset (\lambda^{\mathbf{a}}, \infty)$

$$W^{\pm}(H_a, H_a; M_a E_a(\Lambda))(f_a \otimes \psi^{\mathbf{a}}) = (\Omega^{\pm}(m_a)\hat{\chi}_{\mathbf{a},\Lambda} f_a) \otimes \psi^{\mathbf{a}},$$

where $\mathcal{F}_a \hat{\chi}_{\mathbf{a},\Lambda}\mathcal{F}_a^*$ is multiplication by $\chi_\Lambda(\lambda^{\mathbf{a}} + |\xi_a|^2)$.

Indeed, according to (3.3), (3.4) (applied in the variable x_a), the function $\exp(-iH_a t)(f_a \otimes \psi^{\mathbf{a}})$ equals

$$\exp(i(4t)^{-1}|x_a|^2 - i\lambda^{\mathbf{a}}t)(2it)^{-d_a/2}\hat{f}_a(x_a/(2t))\psi^{\mathbf{a}}(x^a), \tag{13.23}$$

up to an error whose norm is $o(1)$ as $|t| \to \infty$. It is easy to see that (13.23) tends to zero as $|t| \to \infty$ off any conical neighbourhood of X_a. Therefore, by calculation of

$$M_a \exp(-iH_a t)(f_a \otimes \psi^{\mathbf{a}}),$$

we can neglect values of the function $m_a(x)$ everywhere except the neighbourhood of X_a where it depends on x_a only (for $|x| \geq 1$) and replace M_a by the operator

$$\tilde{M}_a = -i\langle \nabla m_a(x_a), \nabla_a \rangle - i\langle \nabla_a, \nabla m_a(x_a) \rangle.$$

This yields

$$W^{\pm}(H_a, H_a; M_a E_a(\Lambda))(f_a \otimes \psi^{\mathbf{a}}) = (W^{\pm}(K_a, K_a; \tilde{M}_a E_{K_a}(\Lambda_{\mathbf{a}}))f_a) \otimes \psi^{\mathbf{a}},$$

where $K_a = -\Delta_{x_a}$, $\Lambda_{\mathbf{a}} = \Lambda - \lambda^{\mathbf{a}}$, and we can apply (13.21) in the variable x_a.

It follows from Lemma 13.7 that if $g_a^{\pm} = \Omega^{\pm}(m_a)\hat{\chi}_{\mathbf{a},\Lambda} f_a$, then

$$\exp(iHt)\exp(-iH_a t)(g_a^{\pm} \otimes \psi^{\mathbf{a}}) = \exp(iHt)M_a \exp(-iH_a t)E_a(\Lambda)(f_a \otimes \psi^{\mathbf{a}}) + o(1)$$

as $t \to \pm\infty$. Therefore the existence of the first wave operator (13.8) implies that the limit $W^{\pm}(H, H_a)(g_a^{\pm} \otimes \psi^{\mathbf{a}})$ also exists. Since, by definition (13.22), the set of elements $g_a^{\pm}$, corresponding to all m_a satisfying the assumptions of Theorem 13.4 and all admissible Λ, is dense in $L_2(X_a)$, the wave operator $W^{\pm}(H, H_a; \mathsf{P}_a)$ exists. Of course these limits are automatically isometric on $\operatorname{Ran} \mathsf{P}_a$.

As we have seen already, the function $\exp(-iH_a t)(f_a \otimes \psi^{\mathbf{a}})$ "lives" in a neighbourhood of the collision plane X_a. Moreover, it follows from its asymptotics (13.23) that this function tends to zero in neighbourhoods of all $X_b \subset X_a$, $X_b \neq X_a$, if $\hat{f}_a(\xi_a) = 0$ in these neighbourhoods. Thus functions $\exp(-iH_a t)(f_a \otimes \psi^{\mathbf{a}})$, for different a, are localized as $|t| \to \infty$ in different regions of the configuration space. This ensures the orthogonality of the subspaces $\operatorname{Ran} \mathsf{P}_a$.

Since the wave operators $W_a^\pm = W^\pm(H, H_a; \mathsf{P}_a)$ exist, it follows from (11.21) that $f = \sum_a W_a^\pm f_a^\pm$. This concludes the proof of equality (11.17) and hence of Theorem 11.1.

13.4. Dispersive Hamiltonians. Let us, finally, make some remarks about dispersive Hamiltonians introduced in subsection 11.5. The first step to construction of scattering theory is a proof of the Mourre estimate (4.9) which, in particular, requires a correct definition of thresholds. Some results in this direction can be found in [38, 56]. The Mourre estimate automatically implies the absence of the singular continuous spectrum and the limiting absorption principle. The next step is to obtain radiation estimates of the same type as those of Theorem 13.2. Such estimates would imply asymptotic completeness, but their proof seems to meet with some obstacles.

14. THE SCATTERING MATRIX AND EIGENFUNCTIONS FOR MULTIPARTICLE SYSTEMS

14.1. Basic definitions. Theorem 11.1 bis implies, of course, that the scattering operator

$$\mathbf{S} = W^{+}(H, \hat{H}; \hat{J})^{*} W^{-}(H, \hat{H}; \hat{J})$$

is unitary in the space (11.18) and commutes with the operator $\hat{H}$. Clearly, $\mathbf{S}$ is the matrix operator with components

$$\mathbf{S}_{\mathbf{ab}} = (W_{\mathbf{a}}^{+})^{*} W_{\mathbf{b}}^{-} : L_2(X_b) \to L_2(X_a), \quad W_{\mathbf{a}}^{\pm} = W^{\pm}(H, K_{\mathbf{a}}; J^{\mathbf{a}}),$$

$\mathbf{a} = \{a, n\}$, $\mathbf{b} = \{b, m\}$, $K_{\mathbf{a}} = K_a + \lambda^{\mathbf{a}}$ and $\mathbf{S}_{\mathbf{ab}} K_{\mathbf{b}} = K_{\mathbf{a}} \mathbf{S}_{\mathbf{ab}}$.

To define the corresponding component $S_{\mathbf{ab}}(\lambda)$ of the scattering matrix, we diagonalize the operator $K_{\mathbf{a}}$. Set (cf. (1.21), (2.12))

$$(\Gamma_{\mathbf{a}}(\lambda) f)(\omega_a) = (\kappa^{\mathbf{a}})^{-1+d_a/2} 2^{-1/2} (\mathcal{F}_a f)(\kappa^{\mathbf{a}} \omega_a), \quad \kappa^{\mathbf{a}} = (\lambda - \lambda^{\mathbf{a}})^{1/2}, \tag{14.1}$$

where $\mathcal{F}_a$ is the Fourier transform in the space $L_2(X_a)$ and

$$(F_{\mathbf{a}} f)(\lambda; \omega_a) = (\Gamma_{\mathbf{a}}(\lambda) f)(\omega_a).$$

Then $F_{\mathbf{a}} : L_2(X_a) \to L_2([\lambda^{\mathbf{a}}, \infty); \mathfrak{N}_a)$, $\mathfrak{N}_a = L_2(\mathbb{S}^{d_a - 1})$, $d_a = \dim X_a$, is unitary and $(F_{\mathbf{a}} K_{\mathbf{a}} f)(\lambda) = \lambda (K_{\mathbf{a}} f)(\lambda)$. Therefore $F_{\mathbf{a}} \mathbf{S}_{\mathbf{ab}} F_{\mathbf{b}}^{*}$ acts as multiplication by the operator-function

$$S_{\mathbf{ab}}(\lambda) : L_2(\mathbb{S}^{d_b - 1}) \to L_2(\mathbb{S}^{d_a - 1})$$

defined for almost all $\lambda \in [\max\{\lambda^{\mathbf{a}}, \lambda^{\mathbf{b}}\}, \infty)$. The corresponding "matrix" operator with components $S_{\mathbf{ab}}(\lambda)$ parametrized by all $\mathbf{a}$ and $\mathbf{b}$ gives the scattering matrix $S(\lambda)$ for the pair $\hat{H}, H$.

14.2. Three-particle one-channel systems. In the three-particle case the structure of the scattering matrix $S(\lambda)$ can be described in a rather detailed way. We proceed here from the results of Section 12. Thus the Hamiltonian H is defined by formulas (11.1), (11.2) where $N = 3$, dimensions of particles $k \geq 3$ and the center of mass is fixed.

Let us start with the case (see subsection 12.2) where all pair operators $H^{\alpha} \geq 0$, and hence the wave operators $W^{\pm}(H, H_0)$ exist and are complete. This implies that the scattering operator $\mathbf{S}$ is unitary in the space $\mathcal{H} = L_2(\mathbb{R}^{2k})$, and the scattering matrix $S(\lambda)$ is defined for almost all $\lambda > 0$ and is unitary in the space $L_2(\mathbb{S}^{2k-1})$. Of course, the diagonalization of the operator H_0 is performed by formulas (1.21), (2.12) where $d = 2k$, only the masses m_1, m_2, m_3 should be taken into account there as numerical coefficients. Let us show that, under the assumptions of Theorem 12.7, the pair H_0, H satisfies the conditions of Proposition 7.2 with $J = I$.

Lemma 14.1 *The operators*

$$\langle x^\alpha\rangle^{-l}\Gamma_0^*(\lambda) : L_2(\mathbf{S}^{2k-1}) \to L_2(\mathbf{R}^{2k}), \quad l > 1, \tag{14.2}$$

are bounded and (as well as their adjoints $\Gamma_0(\lambda)\langle x^\alpha\rangle^{-l}$*) are strongly continuous in* $\lambda > 0$.

In view of (7.4) we have that

$$\langle x^\alpha\rangle^{-l}\Gamma_0^*(\lambda)\Gamma_0(\lambda)\langle x^\beta\rangle^{-l} = \langle x^\alpha\rangle^{-l} dE_0(\lambda)/d\lambda\langle x^\beta\rangle^{-l}, \quad l > 1. \tag{14.3}$$

If $\alpha = \beta$, then, by (1.2) and Proposition 12.2, this operator is bounded and hence (14.2) is also well-defined as a bounded operator. Another possibility to see this fact is to apply Theorem 3.4 (or rather its minor modification) for the case $l_1 = 0, l^1 = l$ to the operator $\Gamma_0(\lambda)\langle x^\alpha\rangle^{-l}$. To prove strong continuity of (14.2) we use an obvious identity

$$(\langle x^\alpha\rangle^{-l}\Gamma_0^*(\lambda)f)(x) = \lambda^{(k-1)/2}(1+|x^\alpha|^2)^{-l/2}(1+\lambda|x^\alpha|^2)^{l/2}(\langle x^\alpha\rangle^{-l}\Gamma_0^*(1)f)(\lambda^{1/2}x)$$

and take into account that the group of dilations is strongly continuous.

Recall also that, by Proposition 12.6, operator-functions $\langle x^\alpha\rangle^{-l}R(z)\langle x^\beta\rangle^{-l}$, $l > 1$, are norm-continuous off the exceptional set $\mathcal{N}$ (closed and of measure zero). Hence it follows from Proposition 7.2 that $S(\lambda)$ is given by formula (8.1), that is

$$S(\lambda) = I - 2\pi i\Gamma_0(\lambda)\Big(\sum_\alpha V^\alpha - \sum_{\alpha,\beta} V^\alpha R(\lambda + i0)V^\beta\Big)\Gamma_0^*(\lambda). \tag{14.4}$$

Thus we obtain

Theorem 14.2 *Under the assumptions of Theorem* 12.7, *the scattering matrix* $S(\lambda)$ *is given by formula* (14.4). *It is a strongly continuous function of* $\lambda > 0$.

We emphasize that formally representation (14.4) is the same as (8.1) in the two-particle problem.

Representation (14.4) allows us to describe the structure of the scattering matrix up to compact terms. Since, by Proposition 12.6, the operators $\langle x^\alpha\rangle^{-l}R(\lambda + i0)\langle x^\beta\rangle^{-l}$, $\alpha \neq \beta$, $l > 1$, are compact in $\mathcal{H}$, all operators

$$\Gamma_0(\lambda)V^\alpha R(\lambda + i0)V^\beta\Gamma_0^*(\lambda), \quad \alpha \neq \beta, \tag{14.5}$$

are compact in the space $L_2(\mathbf{S}^{2k-1})$. Moreover, using the resolvent equation (12.3), we see that operators

$$\Gamma_0(\lambda)V^\alpha(R(\lambda+i0)-R_\alpha(\lambda+i0))V^\alpha\Gamma_0^*(\lambda) = \Gamma_0(\lambda)V^\alpha R(\lambda+i0)\sum_{\beta\neq\alpha} V^\beta R_\alpha(\lambda+i0)V^\alpha\Gamma_0^*(\lambda)$$

are also compact. Thus

$$S(\lambda) - I + 2\pi i\sum_\alpha \Gamma_0(\lambda)(V^\alpha - V^\alpha R_\alpha(\lambda + i0)V^\alpha)\Gamma_0^*(\lambda) \in \mathfrak{S}_\infty. \tag{14.6}$$

Let us now introduce the scattering matrices

$$S_\alpha(\lambda) = S(\lambda; H_\alpha, H_0)$$

for the Hamiltonians (12.2) containing only one pair interaction V^α. Applying representation (14.4) to each $S_\alpha(\lambda)$ and taking into account (14.6), we see that

$$S(\lambda) - I - \sum_\alpha (S_\alpha(\lambda) - I) \in \mathfrak{S}_\infty. \tag{14.7}$$

Moreover, it follows from Proposition 12.6 that operators (14.3) for $\alpha \neq \beta$ are compact in $\mathcal{H}$ which implies that

$$(S_\alpha(\lambda) - I)(S_\beta(\lambda) - I) \in \mathfrak{S}_\infty, \quad \alpha \neq \beta.$$

Therefore relation (14.7) can be rewritten as

$$S(\lambda) = S_{12}(\lambda)S_{23}(\lambda)S_{31}(\lambda)\tilde{S}(\lambda), \quad \text{where} \quad \tilde{S}(\lambda) - I \in \mathfrak{S}_\infty. \tag{14.8}$$

Let us summarize the results obtained.

Theorem 14.3 *Under the assumptions of Theorem* 12.7, *the scattering matrix* $S(\lambda)$ *satisfies relations* (14.7) *and* (14.8).

Factorization (14.8) can be viewed as a non-trivial generalization of the following simple fact. If one of the masses, say m_1, is infinite and $V^{(23)} = 0$, then the Hamiltonian $H = H_0 + V^{(12)} + V^{(31)}$ admits the separation of variables in coordinates x^{12}, x^{31}. In this case the scattering matrix is factorized as $S(\lambda) = S_{12}(\lambda)S_{31}(\lambda)$.

14.3. Three-particle multichannel systems. Let us consider now the case where all three operators H^α have exactly one negative eigenvalue λ^α. Then the scattering matrix $S(\lambda)$ acts in the space

$$L_2(\mathbb{S}^{2k-1}) \oplus L_2(\mathbb{S}^{k-1}) \oplus L_2(\mathbb{S}^{k-1}) \oplus L_2(\mathbb{S}^{k-1})$$

and is determined by its components

$$S_{00}(\lambda) : L_2(\mathbb{S}^{2k-1}) \to L_2(\mathbb{S}^{2k-1}), \quad S_{0\alpha}(\lambda) : L_2(\mathbb{S}^{k-1}) \to L_2(\mathbb{S}^{2k-1}),$$
$$S_{\alpha 0}(\lambda) : L_2(\mathbb{S}^{2k-1}) \to L_2(\mathbb{S}^{k-1}), \quad S_{\alpha\beta}(\lambda) : L_2(\mathbb{S}^{k-1}) \to L_2(\mathbb{S}^{k-1}).$$

The first three operators are defined for $\lambda > 0$, and the last one, $S_{\alpha\beta}(\lambda)$, for $\lambda > \max\{\lambda^\alpha, \lambda^\beta\}$. It can easily be checked that the triple (see subsection 12.4) $\hat{H}$, H, $\hat{\mathbf{J}}$ satisfies the conditions of Proposition 7.2 which leads to a stationary representation of $S(\lambda)$ and hence of its components. Recall that the operators $J^\alpha, \mathbf{J}^0$ and Γ_α, Γ_0 were defined by formulas (12.19), (12.26) and (14.1), respectively. According to (12.31) $W^\pm(\hat{H}, \hat{H}; \hat{\mathbf{J}}^*\hat{\mathbf{J}}) = I$. Let the operators $\mathbf{V}_0$ and $\mathbf{V}_\alpha$ be given by equalities (12.27) and (12.28). Then it follows from general formula (7.11) that

$$\left.\begin{aligned}
S_{00}(\lambda) &= I - 2\pi i\Gamma_0(\lambda)(\mathbf{J}^0\mathbf{V}_0 - \mathbf{V}_0^* R(\lambda + i0)\mathbf{V}_0)\Gamma_0^*(\lambda), \\
S_{0\alpha}(\lambda) &= -2\pi i\Gamma_0(\lambda)(\mathbf{J}^0\mathbf{V}_\alpha - \mathbf{V}_0^* R(\lambda + i0)\mathbf{V}_\alpha)\Gamma_\alpha^*(\lambda), \\
S_{\alpha 0}(\lambda) &= -2\pi i\Gamma_\alpha(\lambda)((J^\alpha)^*\mathbf{V}_0 - \mathbf{V}_\alpha^* R(\lambda + i0)\mathbf{V}_0)\Gamma_0^*(\lambda), \\
S_{\alpha\beta}(\lambda) &= \delta_{\alpha\beta}I - 2\pi i\Gamma_\alpha(\lambda)((J^\alpha)^*\mathbf{V}_\beta - \mathbf{V}_\alpha^* R(\lambda + i0)\mathbf{V}_\beta)\Gamma_\beta^*(\lambda),
\end{aligned}\right\} \tag{14.9}$$

where $\delta_{\alpha\alpha} = 1$ and $\delta_{\alpha\beta} = 0$ if $\alpha \neq \beta$. If instead of (7.11) formula (7.12) is used, then the operators $\mathbf{J}^0\mathbf{V}_0$, $\mathbf{J}^0\mathbf{V}_\alpha$, $(J^\alpha)^*\mathbf{V}_0$, $(J^\alpha)^*\mathbf{V}_\beta$ in the expressions (14.9) for S_{00}, $S_{0\alpha}$, $S_{\alpha 0}$, $S_{\alpha\beta}$ will be replaced by $\mathbf{V}_0^*\mathbf{J}^0$, $\mathbf{V}_0^* J^\alpha$, $\mathbf{V}_\alpha^*\mathbf{J}^0$, $\mathbf{V}_\alpha^* J^\beta$, respectively.

Representations (14.9) show that the operators $S_{0\alpha}$ and $S_{\alpha 0}$ are compact. The components $S_{\alpha\beta}(\lambda)$ have essentially the two-particle structure; in particular, the operators $S_{\alpha\beta}(\lambda) - \delta_{\alpha\beta} I$ are compact.

Thus Theorems 14.2 and 14.3 can naturally be extended to the case where the negative spectra of the operators H^α are not empty. We note that the result of Theorem 14.3 was obtained by R. Newton [123]. The proof given above is borrowed from [155].

14.4. A stationary representation of the N-particle scattering matrix. Following [168], we here consider $S(\lambda)$ under the general assumptions of Section 11. To that end we need to reinforce Theorems 13.1, 13.2 used in Section 13 for the proof of the existence and completeness of wave operators. Below Λ is a compact interval not containing eigenvalues and thresholds of the operator H.

Theorem 14.4 *Let the assumptions of Theorem* 13.1 *hold, let* $\operatorname{Re} z \in \Lambda$, $\pm \operatorname{Im} z \geq 0$, $l > 1/2$ *and let* $G(\mathbf{x}_a, \varepsilon)$ *be operator* (13.1). *Then, for all* $\mathbf{x}_a \in Z_a \cap \mathbb{S}^{d-1}$, *all* $\mathbf{x}_b \in Z_b \cap \mathbb{S}^{d-1}$ *and* ε *small enough, the operator-functions*

$$\langle x \rangle^{-l} R(z) \langle x \rangle^{-l}, \quad \langle x \rangle^{-l} R(z) G^*(\mathbf{x}_a, \varepsilon), \quad G(\mathbf{x}_a, \varepsilon) R(z) G^*(\mathbf{x}_b, \varepsilon) \tag{14.10}$$

are continuous in z *in the topologies of the norm, strong and weak, respectively. Moreover, the operators* $\langle x \rangle^{-l} R(\lambda \pm i0) \langle x \rangle^{-l}$ *and* $\langle x \rangle^{-l} R(\lambda \pm i0) G^*(\mathbf{x}_a, \varepsilon)$ *are compact.*

The result on the first function (14.10) is a consequence of the Mourre estimate (4.9) (see references to Theorems 13.1). We refer to [168] for the proof of the uniform boundedness of the third function (14.10) and hence of its weak continuity.

Here we only consider the operators $\langle x \rangle^{-l} R(z) G^*$, where $G = G(\mathbf{x}_a, \varepsilon)$. Applying estimate (13.7) to elements $u = R(z) f$, $z = \lambda + i\varepsilon$, and taking into account that

$$([H, M] R(z) f, R(z) f) = (M R(z) f, f) - (M f, R(z) f) - 2i\varepsilon (M R(z) f, R(z) f),$$

we obtain the bound

$$||G R(z) f||^2 \leq C(|(R(z) f, M f)| + \varepsilon |(M R(z) f, R(z) f)| + ||\langle x \rangle^{-l} (H_0 + I)^{1/2} R(z) f||^2). \tag{14.11}$$

Using the Hilbert identity

$$R(z) = R(z_0) + (z - z_0) R(z_0) R(z), \quad \operatorname{Im} z_0 \neq 0, \tag{14.12}$$

we see that

$$|(M R(z) f, R(z) f)| \leq ||M R(z_0) f|| \, ||R(z) f|| + |z - z_0| \, ||M R(z_0)|| \, ||R(z) f||^2.$$

Let us now set $f = \langle x \rangle^{-l} g$ and denote

$$\mathcal{B}(z) = \langle x \rangle^{-l} (H_0 + I)^{1/2} R(z) \langle x \rangle^{-l}.$$

Then

$$|(R(z)\langle x\rangle^{-l}g, M\langle x\rangle^{-l}g)| \leq C||\mathcal{B}(z)g||\,||g||,$$
$$\varepsilon||R(z)\langle x\rangle^{-l}g||^2 = \operatorname{Im}(\langle x\rangle^{-l}R(z)\langle x\rangle^{-l}g, g)$$

and (14.11) implies that

$$||GR(z)\langle x\rangle^{-l}g||^2 \leq C(\varepsilon||R(z)\langle x\rangle^{-l}||\,||g||^2 + ||\mathcal{B}(z)g||\,||g|| + ||\mathcal{B}(z)g||^2). \quad (14.13)$$

Using again (14.12) we see that the operator-function $\mathcal{B}(z)$ is norm-continuous, and hence its boundary values

$$\mathcal{B}(\lambda \pm i0) \in \mathfrak{S}_\infty. \quad (14.14)$$

Since $||R(z)\langle x\rangle^{-l}|| = 0(\varepsilon^{-1/2})$ by H-smoothness of the operator $\langle x\rangle^{-l}$, (14.13) yields that

$$||GR(z)\langle x\rangle^{-l}|| \leq C \quad (14.15)$$

and

$$|(GR(z)\langle x\rangle^{-l}g, h)| \leq C(\varepsilon^{1/4}||g|| + ||\mathcal{B}(z)g||^{1/2}\,||g||^{1/2} + ||\mathcal{B}(z)g||)||h||. \quad (14.16)$$

Uniform estimate (14.15) together with continuity in norm of $\mathcal{B}(z)$ imply that the operator-function $GR(z)\langle x\rangle^{-l}$ is weakly continuous in z. Passing to the limit $\varepsilon \to 0$ in inequality (14.16) where $z = \lambda + i\varepsilon$, we obtain that

$$||GR(\lambda \pm i0)\langle x\rangle^{-l}g||^2 \leq C(||\mathcal{B}(\lambda \pm i0)g||\,||g|| + ||\mathcal{B}(\lambda \pm i0)g||^2).$$

This estimate, together with (14.14), ensures that the operators $GR(\lambda \pm i0)\langle x\rangle^{-l}$ are compact. The strong continuity of $\langle x\rangle^{-l}R(\lambda \pm i0)G$ follows again from (14.15).

The following assertion complements Proposition 9.3.

Proposition 14.5 *For any* $\mathbf{a}$, *a point* $\mathbf{x}_b \in Z_b \cap \mathbb{S}^{d-1}$ *and* ε *small enough, the operators* $G(\mathbf{x}_b, \varepsilon)J^{\mathbf{a}}\Gamma_{\mathbf{a}}^*(\lambda) : L_2(\mathbb{S}^{d_a-1}) \to \mathcal{H}$ *are bounded and (as well as their adjoints) are strongly continuous in* $\lambda > \lambda^{\mathbf{a}}$.

Let us return to the scattering matrix. Unfortunately the triple $\hat{H}, H, \hat{J}$ does not satisfy the assumptions of Proposition 7.2. Therefore we first apply Proposition 7.2 to the triple $\hat{H}, H, \tilde{J}$ with an auxiliary identification $\tilde{J}$ defined in terms of the operators M_a (see subsection 13.2). This gives a stationary representation for the scattering matrix $\tilde{S}(\lambda) = S(\lambda; H, \hat{H}; \tilde{J})$ corresponding to $\tilde{J}$. Finally we find a connection between the auxiliary $\tilde{S}(\lambda)$ and the physical $S(\lambda) = S(\lambda; H, \hat{H}; \hat{J})$ scattering matrices.

Let us define the identification $\tilde{J}$ by the relation (cf. (11.20))

$$\tilde{J} = \sum_{\mathbf{a}} \tilde{J}^{\mathbf{a}}, \quad \tilde{J}^{\mathbf{a}} = M_a J^{\mathbf{a}},$$

where M_a is differential operator (4.14) with the generating function m_a satisfying assumptions $1^0 - 3^0$ of subsection 13.2. Then the "effective perturbation"

$$\tilde{T}_{\mathbf{a}} = H\tilde{J}^{\mathbf{a}} - \tilde{J}^{\mathbf{a}}K_{\mathbf{a}} \quad (14.17)$$

for the triple $K_{\mathbf{a}}, H, \tilde{J}^{\mathbf{a}}$ equals $\tilde{T}_{\mathbf{a}} = T_a J^{\mathbf{a}}$ where T_a is operator (13.10).

Let us set

$$\left.\begin{aligned} \mathcal{Q}^{+}_{\mathbf{a},\mathbf{b}}(\lambda) &= \Gamma_{\mathbf{a}}(\lambda)(J^{\mathbf{a}})^* M_a T_b J^{\mathbf{b}} \Gamma^*_{\mathbf{b}}(\lambda), \\ \mathcal{Q}^{-}_{\mathbf{a},\mathbf{b}}(\lambda) &= \Gamma_{\mathbf{a}}(\lambda)(J^{\mathbf{a}})^* T^*_a M_b J^{\mathbf{b}} \Gamma^*_{\mathbf{b}}(\lambda), \end{aligned}\right\} \tag{14.18}$$

$$\mathbf{Q}_{\mathbf{a},\mathbf{b}}(\lambda) = \Gamma_{\mathbf{a}}(\lambda)(J^{\mathbf{a}})^* T^*_a R(\lambda + i0) T_b J^{\mathbf{b}} \Gamma^*_{\mathbf{b}}(\lambda) \tag{14.19}$$

and

$$Q^{\pm}_{\mathbf{a},\mathbf{b}}(\lambda) = \mathcal{Q}^{\pm}_{\mathbf{a},\mathbf{b}}(\lambda) - \mathbf{Q}_{\mathbf{a},\mathbf{b}}(\lambda). \tag{14.20}$$

Note that representation (13.10) together with Theorem 14.4 and Proposition 14.5 allow us to rewrite the right-hand sides of (14.18), (14.19) as combinations of bounded operators. This gives a precise sense to these definitions. We emphasize that operators (14.18) are strongly and the operator (14.19) is weakly continuous in λ.

Consider also the auxiliary wave operators

$$\mathbf{w}^{\pm}_{\mathbf{a},\mathbf{b}} = W^{\pm}(K_{\mathbf{a}}, K_{\mathbf{b}}; E_{K_{\mathbf{a}}}(\Lambda)(J^{\mathbf{a}})^* M_a M_b J^{\mathbf{b}} E_{K_{\mathbf{b}}}(\Lambda)).$$

Equality $\mathbf{w}^{\pm}_{\mathbf{a},\mathbf{b}} = 0$ for $\mathbf{a} \neq \mathbf{b}$ is equivalent to the orthogonality of channels. In the case $\mathbf{a} = \mathbf{b}$, it follows from Lemma 13.7 that

$$\mathbf{w}^{\pm}_{\mathbf{a},\mathbf{a}} = \boldsymbol{\Omega}^{\pm}(m_a)^2 \hat{\chi}_{\mathbf{a},\Lambda}$$

with the operator $\boldsymbol{\Omega}^{\pm}(m_a)$ defined by (13.22).

Proposition 7.2 implies now the following

Proposition 14.6 *Let assumption* (11.15) *be fulfilled for all pair potentials V^α. Then the components $\tilde{S}_{\mathbf{a},\mathbf{b}}(\lambda) : L_2(\mathbb{S}^{d_b-1}) \to L_2(\mathbb{S}^{d_a-1})$ of the scattering matrix $\tilde{S}(\lambda)$ satisfy the following two equalities*

$$\tilde{S}_{\mathbf{a},\mathbf{b}}(\lambda) = 4m^{\pm}_{\mathbf{a}}(\lambda)^2 \delta_{\mathbf{a},\mathbf{b}} - 2\pi i Q^{\pm}_{\mathbf{a},\mathbf{b}}(\lambda), \quad \lambda \in \Lambda, \tag{14.21}$$

where $m^{\pm}_{\mathbf{a}}(\lambda)$ is multiplication by $\kappa^{\mathbf{a}} m_a(\pm\omega_a)$, $\delta_{\mathbf{a},\mathbf{a}} = 1$, $\delta_{\mathbf{a},\mathbf{b}} = 0$ if $\mathbf{a} \neq \mathbf{b}$ and $\kappa^{\mathbf{a}}$ is defined in (14.1).

Diagonal components $\tilde{S}_{\mathbf{a},\mathbf{a}}(\lambda)$ admit another representation which looks slightly more convenient than (14.21). Indeed, let us apply Proposition 14.6 to the triple $K_{\mathbf{a}}, H_a, \tilde{J}^{\mathbf{a}}$. Then the corresponding scattering matrix $\tilde{S}^{(0)}_{\mathbf{a},\mathbf{a}}(\lambda)$ satisfies (14.21) if T_a and $R(z)$ in (14.19), (14.20) are replaced by $[H_a, M_a]$ and $R_a(z) = R_{H_a}(z)$, respectively. On the other hand, it follows from Lemma 13.7 that

$$\tilde{S}^{(0)}_{\mathbf{a},\mathbf{a}}(\lambda) = -4m^{+}_{\mathbf{a}}(\lambda) m^{-}_{\mathbf{a}}(\lambda).$$

Therefore comparing (14.21) for $\tilde{S}_{\mathbf{a},\mathbf{a}}(\lambda)$ and $\tilde{S}^{(0)}_{\mathbf{a},\mathbf{a}}(\lambda)$, we find that

$$\tilde{S}_{\mathbf{a},\mathbf{a}}(\lambda) = -4m^{+}_{\mathbf{a}}(\lambda) m^{-}_{\mathbf{a}}(\lambda) - 2\pi i \Gamma_{\mathbf{a}}(\lambda)(J^{\mathbf{a}})^* \Pi_a(\lambda) J^{\mathbf{a}} \Gamma^*_{\mathbf{a}}(\lambda), \quad \lambda \in \Lambda, \tag{14.22}$$

where

$$\Pi_a(\lambda) = M_a(V - V^a)M_a - T_a^* R(\lambda + i0)T_a + [M_a, H_a]R_a(\lambda + i0)[H_a, M_a]. \quad (14.23)$$

The operator $S_{\mathbf{a},\mathbf{b}}(\lambda)$ can (at least formally) be represented by its integral kernel parametrized by angular variables $\omega_a \in \mathbb{S}^{d_a-1}$ and $\omega_b \in \mathbb{S}^{d_b-1}$. As it is physically natural to expect, these kernels acquire additional singularities as ω_a approaches some plane $X_{a'} \subset X_a$, $X_{a'} \neq X_a$ or ω_b approaches some plane $X_{b'} \subset X_b$, $X_{b'} \neq X_b$. We concentrate here on the simplest case where $\omega_a \in Z_a$ and $\omega_b \in Z_b$, so that they do not belong to all such planes $X_{a'}$ and $X_{b'}$, respectively.

To derive representations for components $S_{\mathbf{a},\mathbf{b}}(\lambda)$ of the scattering matrix $S(\lambda) = S(\lambda; H, \hat{H}, \hat{J})$ with physical identification (11.20) we find a relation connecting $\tilde{S}(\lambda)$ and $S(\lambda)$. According to Lemma 13.7,

$$W^{\pm}(H_a, K_{\mathbf{a}}; \tilde{J}^{\mathbf{a}} E_{K_{\mathbf{a}}}(\Lambda)) = J^{\mathbf{a}} \Omega^{\pm}(m_a)\hat{\chi}_{\mathbf{a},\Lambda}.$$

Therefore, by the multiplication theorem for wave operators,

$$\begin{aligned} W^{\pm}(H, K_{\mathbf{a}}; \tilde{J}^{\mathbf{a}} E_{K_{\mathbf{a}}}(\Lambda)) &= W^{\pm}(H, H_a)W^{\pm}(H_a, K_{\mathbf{a}}; \tilde{J}^{\mathbf{a}} E_{K_{\mathbf{a}}}(\Lambda)) \\ &= W^{\pm}(H, K_{\mathbf{a}}; J^{\mathbf{a}})\Omega^{\pm}(m_a)\hat{\chi}_{\mathbf{a},\Lambda}. \end{aligned} \quad (14.24)$$

This equality gives a relation between the scattering operators

$$\hat{\chi}_{\mathbf{a},\Lambda}\Omega^{+}(m_a)\mathbf{S}_{\mathbf{a},\mathbf{b}}\Omega^{-}(m_b)\hat{\chi}_{\mathbf{b},\Lambda} = \tilde{\mathbf{S}}_{\mathbf{a},\mathbf{b}} \quad (14.25)$$

corresponding to the identifications $\hat{J}$ and $\tilde{J}E_{\hat{H}}(\Lambda)$. Taking into account (13.22) we rewrite (14.25) in terms of the scattering matrices as

$$-4m_{\mathbf{a}}^{+}(\lambda)S_{\mathbf{a},\mathbf{b}}(\lambda)m_{\mathbf{b}}^{-}(\lambda) = \tilde{S}_{\mathbf{a},\mathbf{b}}(\lambda), \quad \lambda \in \Lambda. \quad (14.26)$$

Let us now obtain a representation for the sesquilinear form of $S_{\mathbf{a},\mathbf{b}}(\lambda)$. Suppose that $\operatorname{supp} f_{\mathbf{a}} \subset Z_a \cap \mathbb{S}^{d-1}$ (of course $\operatorname{supp} f_{\mathbf{a}} \subset \mathbb{S}^{d_a-1}$) and $m_a(x_a) = 2^{-1}|x_a|$ if $x_a|x_a|^{-1} \in \operatorname{supp} f_{\mathbf{a}}$. Similarly, we require that $\operatorname{supp} f_{\mathbf{b}} \subset Z_b \cap \mathbb{S}^{d_b-1}$ and $m_b(x_b) = 2^{-1}|x_b|$ if $-x_b|x_b|^{-1} \in \operatorname{supp} f_{\mathbf{b}}$. Then it follows from (14.21) and (14.26) that

$$(S_{\mathbf{a},\mathbf{b}}(\lambda)f_{\mathbf{b}}, f_{\mathbf{a}}) = -\delta_{\mathbf{a},\mathbf{b}}(f_{\mathbf{b}}, f_{\mathbf{a}}) + 2\pi i(\kappa^{\mathbf{a}}\kappa^{\mathbf{b}})^{-1}(Q_{\mathbf{a},\mathbf{b}}^{\pm}(\lambda)f_{\mathbf{b}}, f_{\mathbf{a}}), \quad \lambda \in \Lambda. \quad (14.27)$$

The continuity of $(S_{\mathbf{a},\mathbf{b}}(\lambda)f_{\mathbf{b}}, f_{\mathbf{a}})$ on dense sets of elements $f_{\mathbf{a}}, f_{\mathbf{b}}$ together with the bound $\|S_{\mathbf{a},\mathbf{b}}(\lambda)\| \leq 1$ ensure the weak continuity of $S_{\mathbf{a},\mathbf{b}}(\lambda)$. In view of the unitarity of the scattering matrix $S(\lambda)$ the weak convergence can be replaced here by the strong one.

In the case $\mathbf{a} = \mathbf{b}$ we can also proceed from equality (14.22) which gives

$$(S_{\mathbf{a},\mathbf{a}}(\lambda)f_{\mathbf{a}}, f_{\mathbf{a}}) = \|f_{\mathbf{a}}\|^2 - 2\pi i(\lambda - \lambda^{\mathbf{a}})^{-1}(\Pi_a(\lambda)J^{\mathbf{a}}\Gamma_{\mathbf{a}}^*(\lambda)f_{\mathbf{a}}, J^{\mathbf{a}}\Gamma_{\mathbf{a}}^*(\lambda)f_{\mathbf{a}}). \quad (14.28)$$

An advantage of the last representation is that the first term in the right-hand side corresponds to the identity operator (cf. representations (14.9) for $S_{00}(\lambda)$ and $S_{\alpha\alpha}(\lambda)$).

Let us summarize the results obtained.

Theorem 14.7 *Suppose that pair potentials V^α satisfy assumption* (11.15). *Then representations* (14.27) *hold, where the operators $Q^{\pm}_{\mathbf{a},\mathbf{b}}(\lambda)$ are defined by* (14.18) – (14.20). *For any* **a**, **b** *the operator $S_{\mathbf{a},\mathbf{b}}(\lambda)$ is strongly continuous in $\lambda \in \Lambda$. Moreover, $S_{\mathbf{a},\mathbf{a}}(\lambda)$ satisfies* (14.23), (14.28).

Theorem 14.7 shows that the scattering matrix $S(\lambda)$ can be expressed in terms of boundary values of the resolvent $R(z)$ at the point $z = \lambda + i0$. Compared to the two-particle Theorem 8.1 or the three-particle Theorems 14.2 and 14.3, the main drawback of Theorem 14.7 is that operator (14.19) contains the terms

$$G(\mathbf{x}_a, \varepsilon)R(\lambda + i0)G^*(\mathbf{x}_b, \varepsilon)$$

which are not compact. Therefore representation (14.27) for the scattering matrix $S(\lambda)$ does not allow us to describe singularities of its kernel.

We conclude this subsection with

Problem 14.8 *Extend the result of Theorem* 14.3 *to an arbitrary number of particles.*

Actually, even for three particles relations (14.7) and (14.8) are not proven under the only assumption (11.15).

14.5. Eigenfunctions of the continuous spectrum. The technique exposed in the previous subsection allows one [167] to obtain also stationary representations for the wave operators $W^{\pm}(H, K_{\mathbf{a}}; J^{\mathbf{a}})$. This is equivalent (cf. subsection 2.3) to construction of eigenfunctions $\psi_{\mathbf{a}}(x; \lambda, \omega_a)$, $\omega_a \in \mathbb{S}^{d_a-1}$, of the Schrödinger operator smeared over ω_a. Similarly to the study of the scattering matrix, neighbourhoods of collision planes require a separate treatment (see [167]).

Since the triple $\hat{H}, H, \hat{J}$ does not satisfy the assumptions of Proposition 7.1, we are obliged to start with stationary representations of the wave operators

$$W^{\pm}(H, K_{\mathbf{a}}; \tilde{J}^{\mathbf{a}} E_{K_{\mathbf{a}}}(\Lambda)).$$

It follows from Theorem 14.4 and Proposition 14.5 that perturbation (14.17) can be factored into a product of $K_{\mathbf{a}}$ - and H-smooth operators. Therefore Proposition 7.1 can be directly applied to the triple $K_{\mathbf{a}}, H, \tilde{J}^{\mathbf{a}}$. Recall that the operator $\Gamma_{\mathbf{a}}(\lambda)$ was defined by (14.1). In the case considered formulas (7.7) and (7.8) take, respectively, the form

$$\Gamma^{\pm}_{\mathbf{a}}(\lambda) = \Gamma_{\mathbf{a}}(\lambda)(J^{\mathbf{a}})^*(M_a - T^*_a R(\lambda \pm i0)), \quad \lambda \in \Lambda,$$

and

$$(W^{\pm}(H, K_{\mathbf{a}}; \tilde{J}^{\mathbf{a}} E_{K_{\mathbf{a}}}(\Lambda))u_{\mathbf{a}}, u) = \int_\Lambda (\Gamma_{\mathbf{a}}(\lambda)u_{\mathbf{a}}, \Gamma^{\pm}_{\mathbf{a}}(\lambda)u)_{L_2(\mathbb{S}^{d_a-1})} d\lambda. \tag{14.29}$$

For any $l > 1/2$, the operators $\Gamma_{\mathbf{a}}(\lambda) : L_2^{(l)}(X_a) \to L_2(\mathbb{S}^{d_a-1})$ and $\Gamma^{\pm}_{\mathbf{a}}(\lambda) : L_2^{(l)}(\mathbb{R}^d) \to L_2(\mathbb{S}^{d_a-1})$ are bounded. The first of them is continuous in norm and the second is weakly continuous with respect to $\lambda \in \Lambda$. Hence the integrand in (14.29) is a continuous function of λ provided $u_{\mathbf{a}} \in L_2^{(l)}(X_a)$, $u \in L_2^{(l)}(\mathbb{R}^d)$ for some $l > 1/2$.

The next step is to pass to a stationary representation of the physical wave operators $W^{\pm}(H, K_{\mathbf{a}}; J^{\mathbf{a}} E_{K_{\mathbf{a}}}(\Lambda))$. We proceed again from identity (14.24). Let $\hat{g}_{\mathbf{a}} \in C_0^\infty(Z_a)$ and $m_a(x) = 2^{-1}|x_a|$ if $\pm x_a \in \operatorname{supp} \hat{g}_{\mathbf{a}}$. By (13.22),

$$(\mathcal{F}_a \Omega^{\pm}(m_a)\hat{\chi}_{\mathbf{a},\Lambda} g_{\mathbf{a}})(\xi_a) = \pm|\xi_a|\chi_\Lambda(\lambda^{\mathbf{a}} + |\xi_a|^2)\hat{g}_{\mathbf{a}}(\xi_a).$$

Set now $\hat{u}_{\mathbf{a}}(\xi_a) = |\xi_a|\hat{g}_{\mathbf{a}}(\xi_a)$. Applying (14.29) to functions $g_{\mathbf{a}}$, u, we find that

$$(W^{\pm}(H, K_{\mathbf{a}}; J^{\mathbf{a}} E_{K_{\mathbf{a}}}(\Lambda))u_{\mathbf{a}}, u) = \pm \int_\Lambda (\lambda - \lambda^{\mathbf{a}})^{-1/2} (\Gamma_{\mathbf{a}}(\lambda)u_{\mathbf{a}}, \Gamma_{\mathbf{a}}^{\pm}(\lambda)u)_{L_2(\mathbb{S}^{d_a - 1})} d\lambda. \tag{14.30}$$

Now we can formulate the result obtained.

Theorem 14.9 *Suppose that pair potentials* V^α *satisfy assumption* (11.15). *Let* $\hat{u}_{\mathbf{a}} \in C_0^\infty(Z_a)$, *let* m_a *obey conditions* $1^0 - 3^0$ *of subsection* 13.2 *and* $m_a(x) = 2^{-1}|x_a|$ *if* $\pm x_a \in \operatorname{supp} \hat{u}_{\mathbf{a}}$. *Then for any* $u \in L_2^{(l)}(\mathbb{R}^d)$, $l > 1/2$, *representation* (14.30) *holds.*

Note that for any $f_{\mathbf{a}} \in L_2(\mathbb{S}^{d_a - 1})$ such that $\operatorname{supp} f_{\mathbf{a}} \subset Z_a \cap \mathbb{S}^{d-1}$, the functions $\psi = \Gamma_{\mathbf{a}}^{\pm}(\lambda)^* f_{\mathbf{a}}$ satisfy the the Schrödinger equation $H\psi = \lambda\psi$. Similarly to the two-particle case, (14.30) can be considered as an expansion theorem in terms of such solutions.

14.6. A brief review. Here we give a very brief account of results of different authors related to the topics of this section. These results are obtained by quite different methods, and some of them are probably not quite rigorous.

The paper [30] is devoted to a study of singularities of the scattering matrix $S(\lambda)$ for three-particle systems under, roughly speaking, the assumptions of subsections 14.2 or 14.3. According to (14.4) the integral operator $S(\lambda) - I$ contains the terms $S_\alpha(\lambda) - I$ with Dirac-functions in its kernels and terms (14.5) with singular denominators of Cauchy type. One of the results of [30] is a description of all singularities of $S(\lambda)$ in terms of scattering matrices for two-particle subsystems. Compared to Theorem 14.3 this result has a more detailed character, but it is formulated in an essentially more complicated way. In particular, compact terms (14.5) are considered as singular in [30].

From the view-point of resolvent equations, the case of one-dimensional particles is more singular than the case of particles of dimension $k \geq 3$ (this condition was imposed in Section 12 and subsections 14.2 and 14.3). The description of principle singularities of $S(\lambda)$ for three one-dimensional particles, similar to that of [30], was given in [27, 31]. Then these results were extended to an arbitrary number of particles in [28]. For one-dimensional particles a diffraction picture plays an important role.

For many-particle systems the definition of the total cross-section is basically similar to the two-particle case (see subsection 8.5 and especially formula (8.21)). In the papers [7, 47] the finiteness of the total cross-section was verified for an incoming two-cluster channel and an arbitrary outgoing channel. It is of course required in [7, 47] that pair potentials satisfy condition (11.15) with $\rho > (k+1)/2$.

To be more precise, the cross-section is smeared in [7, 47] over the energy λ which is intrinsic for the time-dependent technique used in these papers. A more difficult problem of finiteness of the cross-section for fixed λ seems to remain open. We note however papers [144], [80] where, for an initial two-cluster channel $\mathbf{a}$ and the N-cluster outgoing channel, a regularity of $S_{0,\mathbf{a}}(\omega, \omega_a; \lambda)$ was verified away from the bad directions (collision planes) of ω and ω_a.

Let us now turn to construction of eigenfunctions of the N-particle Schrödinger operator. Similarly to the case $N = 2$, this problem can be formulated on different levels. In the previous subsection we have constructed eigenfunctions $\psi_{\mathbf{a}}(x; \lambda, \omega_a)$ smeared over ω_a. This can be considered as a natural modification of the construction of subsection 2.3. A more difficult problem is to construct $\psi_{\mathbf{a}}(x; \lambda, \omega_a)$ for a fixed ω_a. For $N = 2$ this is possible (see Theorem 3.9) if V satisfies (1.15) for some $\rho > (k+1)/2$; in this case the function $\psi(x; \lambda, \omega)$ can be defined by formula (1.18). For an arbitrary N, $a = 0$ and ω not belonging to the collision planes a construction of $\psi_0(x; \lambda, \omega)$ was given in [64].

Actually, the most interesting problem is to find the asymptotics of $\psi_{\mathbf{a}}(x; \lambda, \omega_a)$ as $|x| \to \infty$. For three three-dimensional particles this problem was studied in [118] and [126] (see also the book [50]). It is important that in this case the asymptotics of $\psi_0(x; \lambda, \omega)$ contains not only the plane and spherical waves (cf. (1.16)) but also Frenel integrals. On the contrary, for the case $a = (i, j)(k)$ (that is for the two cluster decomposition) the asymptotics of $\psi_{\mathbf{a}}(x; \lambda, \omega_a)$ was related in [78, 79] to the scattering matrix $S_{0,\mathbf{a}}(\omega, \omega_a; \lambda)$ by a formula of type (1.16). The case of one-dimensional particles was considered in [31, 28].

We finally mention a recent series of papers by A. Vasy (see e.g. [150] and references therein). His approach seems to rely heavily on the theory of propagation of singularities and thus is out of the scope of the present survey.

15. NEW CHANNELS OF SCATTERING

Here we consider two- and N- particle systems with anysotropic and long-range interactions for which there exist channels of scattering not taken into account in the definition of the asymptotic completeness.

15.1. A general construction. Let us start with the general construction suggested in [157] and refined in [169, 170]. Suppose that $\mathbf{R}^d$ is decomposed into orthogonal sum (3.7), but we do not make any special assumptions about a potential $V(x) = V(x_1, x^1)$. Let us introduce the operator

$$H^1(x_1) = -\Delta_{x^1} + V(x_1, x^1) \tag{15.1}$$

acting on the space $L_2(X^1)$. Suppose that $H^1(x_1)$ has a (negative) eigenvalue $\lambda(x_1)$, and denote by $\psi(x_1, x^1)$ a corresponding normalized eigenfunction. Actually, it suffices to assume that $\psi(x_1, x^1)$ is an approximate eigenfunction, that is

$$-\Delta_{x^1}\psi(x_1, x^1) + V(x_1, x^1)\psi(x_1, x^1) = \lambda(x_1)\psi(x_1, x^1) + Y(x_1, x^1), \tag{15.2}$$

where

$$||Y(x_1)||_{L_2(X^1)} = O(|x_1|^{-1-\varepsilon}) \quad \text{as} \quad |x_1| \to \infty. \tag{15.3}$$

In interesting situations the function $\lambda(x_1)$ tends to zero slower than $|x_1|^{-1}$. Suppose that $\lambda(x_1)$ satisfies condition (4.1). Let us consider it as an "effective" potential energy and associate (see subsection 4.1) to the long-range potential $\lambda(x_1)$ the phase function $\Xi_1(x_1, t)$. Now Ξ_1 satisfies the eikonal equation (4.3) with $V(x)$ replaced by $\lambda(x_1)$, that is

$$\partial\Xi_1/\partial t + |\nabla_{x_1}\Xi_1|^2 + \lambda(x_1) = 0, \tag{15.4}$$

perhaps, up to a term $O(|x_1|^{-1-\varepsilon})$. Let us introduce the operator of a modified free evolution (cf. (4.2)) $U_1(t)$ in the space $L_2(X_1)$,

$$(U_1(t)f_1)(x) = \exp(i\Xi_1(x_1, t))(2it)^{-d_1/2}\hat{f}_1(x_1/(2t)), \tag{15.5}$$

and let an isometric operator $J : L_2(X_1) \to L_2(\mathbf{R}^d)$ be defined by the formula

$$(Jg)(x_1, x^1) = \psi(x_1, x^1)g(x_1), \quad J = J(\psi).$$

Set, as usual, $H = -\Delta + V$. Our goal is to show the existence of the limits

$$w^\pm = \lim_{t\to\pm\infty} \exp(iHt)JU_1(t). \tag{15.6}$$

These operators are isometric on $L_2(X_1)$, and the intertwining property holds:

$$Hw^\pm = w^\pm(-\Delta_{x_1}).$$

If $V(x_1, x^1) = V(x^1)$, then H is the three-particle Hamiltonian with only one non-trivial pair interaction. In this case $\psi(x_1, x^1) = \psi(x^1)$ is an eigenfunction of this pair, λ does not depend on x_1 and $w^\pm$ describes scattering of the third particle on this bound state. In the general case H can be considered as the three-particle Hamiltonian with potential energy depending sufficiently arbitrarily on positions of particles. Then $w^\pm$ describes a channel where a pair of particles is in a bound state depending on the position of the third particle, and the third particle is being scattered by the "effective" potential $\lambda(x_1)$.

The existence of wave operators (15.6) requires rather special assumptions which are naturally formulated in terms of eigenfunctions $\psi(x_1, x^1)$. Typically the asymptotic behaviour of $\psi(x_1, x^1)$ as $\lambda(x_1) \to 0$ has a certain self-similarity:

$$\psi(x_1, x^1) = b(x_1)^{-d^1/2}\Psi(b(x_1)^{-1}x^1) + o(1) \tag{15.7}$$

for some $\Psi \in L_2(X^1)$ and some function $b(x_1) \to \infty$ as $|x_1| \to \infty$. In the examples below $b(x_1) =< x_1 >^\sigma= (1 + |x_1|^2)^{\sigma/2}$, $\sigma > 0$. We prove the existence of wave operators (15.6) if $\sigma < 1/2$ in (15.7) and some conditions on the function Ψ and on the remainder $o(1)$ are fulfilled. On the other hand, simple examples (see subsections 15.2 and 15.3 below) show that (15.7) for $\sigma > 1/2$ does not ensure the existence of $w^\pm$.

Limits (15.6) exist provided

$$\int_{-\infty}^{\infty} ||(i\partial/\partial t - H)U_1(t)f_1||dt < \infty, \quad \hat{f}_1 \in C_0^\infty(\mathbf{R}^{d_1} \setminus \{0\}), \tag{15.8}$$

that is $U_1(t)f_1$ is a reasonably good approximate solution of the time-dependent Schrödinger equation. Unfortunately, the integrand in (15.8) behaves as $|t|^{-1}$ for $|t| \to \infty$, so condition (15.8) is never satisfied. It turns out however that this condition holds for a modified free evolution defined by

$$(\tilde{U}_1(t)f_1)(x_1, x^1) = \exp(i\sigma(4t)^{-1}(x^1)^2)(U_1(t)f_1)(x_1, x^1).$$

This implies existence of limits (15.6) with U_1 replaced by $\tilde{U}_1$. Since moreover

$$\lim_{t\to\pm\infty} ||\tilde{U}_1(t)f_1 - U_1(t)f_1|| = 0, \quad \hat{f}_1 \in C_0^\infty(\mathbf{R}^{d_1} \setminus \{0\}),$$

the wave operators $w^\pm$ also exist.

According to (15.7) functions (15.5) "live" in the region where $|x^1| \leq C|x_1|^\sigma$. This implies that, for the evolution $U_0(t)$ defined by formula (4.2), where the phase function $\Xi(x, t)$ may be arbitrary,

$$\lim_{|t|\to\infty} (U_0(t)f_0, JU_1(t)f_1) = 0, \quad \forall \hat{f}_0 \in C_0^\infty(\mathbf{R}^d \setminus \{0\}),\ \forall \hat{f}_1 \in C_0^\infty(\mathbf{R}^{d_1} \setminus \{0\}). \tag{15.9}$$

15.2. Anysotropic potentials. The simplest class of potentials for which the conditions of the previous subsection are satisfied is determined by the formula

$$V(x_1, x^1) =< x_1 >^{-2\sigma} V_0(< x_1 >^{-\sigma} x^1). \tag{15.10}$$

Suppose that the operator

$$K = -\Delta_{x^1} + V_0(x^1)$$

in the space $L_2(X^1)$ has a (negative) eigenvalue Λ with eigenfunction $\Psi(x^1)$. Then

$$\psi(x_1, x^1) = < x_1 >^{-\sigma d^1/2} \Psi(< x_1 >^{-\sigma} x^1) \tag{15.11}$$

is an eigenfunction of the operator (15.1) corresponding to the eigenvalue

$$\lambda(x_1) = < x_1 >^{-2\sigma} \Lambda.$$

In this case asymptotic relation (15.7) becomes an equality and the function Y in (15.2) equals zero. Let the phase function $\Xi_1(x_1, t)$ be defined as an approximate solution of equation (15.4), and let the unitary family $U_1(t)$ be defined by formula (15.5). Then the wave operators (15.6) exist provided $\sigma < 1/2$ in (15.10).

Clearly, the decay of function (15.10) depends on the direction of x, and, if $\sigma < 1/2$, the short-range assumption (1.15) with $\rho > 1$ is never fulfilled. However if V_0 satisfies the estimate

$$|V_0(x^1)| \leq C(1 + |x^1|)^{-r}, \quad r > (1 - 2\sigma)(1 - \sigma)^{-1}, \tag{15.12}$$

then $V(x) = O(|x|^{-\rho})$, $\rho > 1$, as $|x| \to \infty$ in any closed cone not intersecting the subspaces X_1 and X^1 and hence, by Theorem 3.2, the wave operators $W^\pm = W^\pm(H, H_0)$ exist. By virtue of (15.9) in this case the ranges of the wave operators $W^\pm$ and $w^\pm$ are orthogonal. Let us formulate the precise result obtained in [157].

Theorem 15.1 *Suppose that* (15.10) *holds for* $\sigma < 1/2$ *and that the eigenfunction* Ψ *of the operator* K *satisfies the condition*

$$\int_{X^1} (1 + |x^1|^4) |(D^\kappa \Psi)(x^1)|^2 dx^1 < \infty, \quad |\kappa| \leq 2.$$

Then the wave operators $w^\pm$ *exist. Under additional assumption* (15.12) *the wave operators* $W^\pm$ *also exist, and the subspaces* Ran $w^\pm$, Ran $W^\pm$ *are orthogonal to each other.*

Let us compare Theorems 3.3 and 15.1. Suppose for simplicity that $V_0(x^1) = v_0|x^1|^{-r}$ (if r is small enough the local singularity does not violate self-adjointness of the operators H and K). Then (15.10) reduces to

$$V(x_1, x^1) = v_0 < x_1 >^{-\rho_1} |x^1|^{-\rho^1}, \quad \text{where} \quad \rho_1 = (2 - r)\sigma, \ \rho^1 = r.$$

Since inequality $\sigma < 1/2$ is equivalent to $\rho_1 + 2^{-1}\rho^1 < 1$, Theorem 15.1 implies that condition $\rho_1 + 2^{-1}\rho^1 > 1$ (as well as $\rho^1 + 2^{-1}\rho_1 > 1$) of Theorem 3.3 for the completeness of the wave operators $W^\pm$ is optimal in the class of potentials satisfying (3.8).

On the other hand, if $\sigma \in (1/2, 1]$ in (15.10),

$$r > 2(1 - \sigma)(2 - \sigma)^{-1}$$

in (15.12) and $d_1 > 1$, $d^1 > 1$, then potential (15.10) satisfies assumptions of Theorem 3.3. Therefore the wave operators $W^\pm$ are complete and hence the operators $w^\pm$ do not exist. This shows that the condition $\sigma < 1/2$ imposed in the previous subsection for the proof of the existence of limits (15.6) cannot be improved.

Similarly to Theorem 15.1, one can check that the wave operators $w^\pm$ exist if

$$V(x) = V_1(x_1)V^1(x^1), \quad d^1 = 1,$$

where the nonnegative function $V^1(x^1)$ may have compact support and $V_1(x_1) = -v_1|x_1|^{-\rho_1}$ with $v_1 > 0$ and $\rho_1 \in (0, 1/2)$. Thus, the condition $\rho_1 + d^1/2 > 1$ (and the condition $\rho^1 + d_1/2 > 1$) of Theorem 3.3 for the completeness of the wave operators $W^\pm$ is also optimal.

15.3. A two-particle example. Another class of potentials (see [170], for details) for which the wave operators $w^\pm$ exist is given by a symmetric formula

$$V(x_1, x^1) = v(< x_1 >^q + < x^1 >^q)^{-\rho/q}, \quad \rho \in (0,1),\ q \in (0,2),\ v < 0. \tag{15.13}$$

This function is infinitely differentiable and $V(x) = O(|x|^{-\rho})$ as $|x| \to \infty$. Moreover, the bound (4.1) is fulfilled for arbitrary κ off any conical neighbourhood of the planes X_1 and X^1. This suffices for the existence of the modified wave operator $\mathcal{W}^\pm$. If $q \in [1, 2)$, then (4.1) is fulfilled (uniformly in directions of x) for $|\kappa| = 1$, but it is violated for $|\kappa| = 2$. If $q \in (0,1)$, then (4.1) fails already for $|\kappa| = 1$.

Let us show that approximate eigenfunctions $\psi(x_1, x^1)$ of the operator (15.1) satisfy condition (15.7) with

$$\sigma = (\rho + q)(2 + q)^{-1}. \tag{15.14}$$

In the case $q > 1 - \rho$ we replace potential (15.13) by the first two terms

$$v < x_1 >^{-\rho} + \omega < x_1 >^{-\rho-q} < x^1 >^q, \quad \omega = -vq^{-1}\rho > 0,$$

of its Taylor expansion. The first term here is independent of x^1 and hence contributes only to the eigenvalue $\lambda(x_1)$. We define $\psi(x_1, x^1)$ by formula (15.11) where Ψ is an eigenfunction of the operator $K = -\Delta_{x^1} + \omega|x^1|^q$ (corresponding to an eigenvalue Λ). Then

$$-\Delta_{x^1}\psi(x_1, x^1) + \omega < x_1 >^{-\rho-q} |x^1|^q \psi(x_1, x^1) =< x_1 >^{-2\sigma} \Lambda\psi(x_1, x^1).$$

It follows that equation (15.2) holds with

$$\lambda(x_1) = v < x_1 >^{-\rho} + \Lambda < x_1 >^{-2\sigma}, \quad 2\sigma > \rho, \tag{15.15}$$

and

$$Y(x_1, x^1) = vy(x_1, x^1)\psi(x_1, x^1)$$

where

$$y(x_1, x^1) = (< x_1 >^q + < x^1 >^q)^{-\rho/q} - < x_1 >^{-\rho} + \rho q^{-1} < x_1 >^{-\rho-q} |x^1|^q.$$

It is easy to see that this function Y satisfies condition (15.3). Therefore the wave operators $w^\pm$ exist provided $\sigma < 1/2$ which by virtue of (15.14) gives the condition $q < 2(1 - \rho)$. The orthogonality of the subspaces $\operatorname{Ran} \mathcal{W}^\pm$ and $\operatorname{Ran} w^\pm$ follows again from equation (15.9). Thus we arrive at the following result.

Theorem 15.2 *Let a potential V be defined by* (15.13) *where $1-\rho<q<2(1-\rho)$. Let Λ be any eigenvalue and Ψ be a corresponding eigenfunction of the operator $K=-\Delta_{x^1}+\omega|x^1|^q$ in the space $L_2(X^1)$. Define the function $\psi(x_1,x^1)$ and the "potential" $\lambda(x_1)$ by the equations* (15.11) *and* (15.15) *where $\sigma=(\rho+q)(2+q)^{-1}$. Let $\Xi_1(x_1,t)$ be the phase function associated to the long-range potential $\lambda(x_1)$, and let $U_1(t)$ be defined by* (15.5). *Then the wave operators $w^{\pm}$ exist. The operators $\mathcal{W}^{\pm}$ also exist and the subspaces* $\operatorname{Ran}\mathcal{W}^{\pm}$, $\operatorname{Ran} w^{\pm}$ *are orthogonal to each other.*

The restriction $q>1-\rho$ here is of technical nature. If $q\leq 1-\rho$ we should keep several terms in expansion of function (15.13) in powers of $<x_1>^{-1}<x^1>$ but cut the functions $<x^1>^{qk}$ for sufficiently large $|x^1|$ if $k\geq 2$. A formula for the effective potential $\lambda(x_1)$ is getting more complicated compared to (15.15) although its leading term as $|x_1|\to\infty$ is always $v<x_1>^{-\rho}$. Equality (15.11) for approximate eigenfunctions $\psi(x_1,x^1)$ of the operator $H(x_1)$ also fails but the $L_2(X^1)$-norm of the difference of its left- and right-hand sides tends to zero as $|x_1|\to\infty$. One first proves the existence of the wave operators $w^{\pm}$ corresponding to $\psi(x_1,x^1)$ but then these functions can be replaced in the definition (15.5) of $U_1(t)$ by the right-hand side of (15.11).

We emphasize that for potentials (15.13) there exists a countable set of wave operators $w_n^{\pm}$ corresponding to each eigenvalue Λ_n of the operator K. The subspaces $\operatorname{Ran} w_n^{\pm}$ are, obviously, orthogonal to one another. Furthermore, one can interchange the roles of the variables x_1 and x^1. This gives us a new set of subspaces orthogonal to $\operatorname{Ran} w_n^{\pm}$ (and, of course, to $\operatorname{Ran}\mathcal{W}^{\pm}$).

If $q=2$, then $V(x)=-v<x>^{-\rho}$ and $\sigma=(\rho+2)/4>1/2$. In this case the wave operators $\mathcal{W}^{\pm}$ are complete, and $w^{\pm}$ do not exist. This shows again that condition (15.7) for $\sigma>1/2$ does not guarantee the existence of the wave operators $w^{\pm}$.

Probably, $\mathcal{W}^{\pm}$ are complete also for $\rho\in(0,1)$, $q\in(2(1-\rho),2)$, but seemingly this does not follow from the known results.

15.4. Three one-dimensional particles. Let us now consider the Schrödinger operator, which describes three one-dimensional particles with one of three pair interactions decaying slower than $|x|^{-1/2}$ at infinity. In contrast to (15.13) this pair potential will satisfy condition (4.1) for all κ. Another pair potential may be short-range and the third one will be even zero. Thus, we consider the operator $H=H_0+V$ where $H_0=-\Delta$ and

$$V(x)=V^1(x^1)+V^2(x^2) \tag{15.16}$$

in the space $\mathcal{H}=L_2(\mathbb{R}^2)$. The subspaces X^1 and X^2 of $\mathbb{R}^2$ are one-dimensional and not orthogonal to each other (and of course $X^1\neq X^2$). Then $x^2=\gamma^1x^1-\gamma_1x_1$ where $(\gamma^1)^2+\gamma_1^2=1$, $\gamma^1\neq 0$, $\gamma_1\neq 0$. Set for simplicity $\gamma^1=\gamma_1=2^{-1/2}$. We suppose that $V^1\geq 0$ is a bounded function, $V^1(x^1)=0$ for $x^1\geq 0$ and $V^1(x^1)=v_1|x^1|^{-r}$, $v_1>0$, $r\in(0,2)$, for large negative x^1. In particular, V^1 is a short-range potential if $r>1$. A long-range pair potential V^2 is assumed to satisfy for some $\rho\in(0,1/2)$ one of the two following conditions:

1^0 $V^2(x^2) = v_2|x^2|^{-\rho}$, $v_2 < 0$, for large positive x^2, $x^2 \geq c$,

2^0 $V^2(x^2) = v_2|x^2|^{-\rho}$, $v_2 > 0$, for large negative x^2, $x^2 \leq -c$.

Under these assumptions the modified wave operators $\mathcal{W}_0^\pm$ (corresponding to the channel where all particles are asymptotically free) and $\mathcal{W}_{2,n}^\pm$ (corresponding to scattering on bound states of the Hamiltonian $H^2 = -\Delta_{x^2}+V^2(x^2)$) exist. However, as we shall see below, the asymptotic completeness (11.17) fails, that is

$$\operatorname{Ran}\mathcal{W}_0^\pm \oplus \bigoplus_n \operatorname{Ran}\mathcal{W}_{2,n}^\pm \neq \mathcal{H}_H^{(ac)}.$$

Let Λ be any eigenvalue and Ψ be a corresponding eigenfunction (the Airy function) of the equation

$$-\Psi'' + |v|\rho x^1\Psi = \Lambda\Psi, \quad x^1 \geq 0, \quad \Psi(0) = 0, \quad v = v_2 2^{\rho/2}, \tag{15.17}$$

extended by 0 to $x^1 \leq 0$. Define the function $\psi(x_1, x^1)$ and the "potential" $\lambda(x_1)$ by the equations

$$\psi(x_1, x^1) = |x_1|^{-\sigma/2}\Psi(|x_1|^{-\sigma}x^1), \quad \sigma = (\rho+1)/3,$$

$$\lambda(x_1) = v|x_1|^{-\rho} + \Lambda|x_1|^{-2\sigma}$$

(cf. (15.11) and (15.15)), and let $\Xi_1(x_1, t)$ be the phase function associated to the long-range potential $\lambda(x_1)$, that is Ξ_1 is an (approximate) solution of equation (15.4).

The following result was obtained in [169].

Theorem 15.3 *Under the assumptions above, the wave operator $w^\pm$ defined by equalities* (15.5), (15.6) *exists for any $\hat{f}_1 \in L_2(\mathbb{R}_\mp)$ in the case* 1^0 *and for any $\hat{f}_1 \in L_2(\mathbb{R}_\pm)$ in the case* 2^0. *Moreover, the subspaces* $\operatorname{Ran} w^\pm$, $\operatorname{Ran}\mathcal{W}_0^\pm$ *and* $\operatorname{Ran}\mathcal{W}_{2,n}^\pm$ *are orthogonal.*

We emphasize that in Theorem 15.3 the functions $\hat{f}_1$ are restricted either to positive or negative half-axis. This means that the subspace $\operatorname{Ran} w^\pm$ contains only "one-half" of states compared to Theorems 15.1 or 15.2 for the case $d_1 = 1$.

The proof follows essentially the scheme exposed above. We construct $\psi(x_1, x^1)$ as a bound state in an effective potential well in the variable x^1 created by the potential

$$V(x) = V^1(x^1) + V^2(2^{-1/2}(x^1 - x_1)) \tag{15.18}$$

as $x_1 \to -\infty$ in the case 1^0 (see Figure 15.1) and as $x_1 \to \infty$ in the case 2^0 (see Figure 15.2). In both cases this well is limited from the left (for $x^1 \leq 0$) by the barrier $V^1(x^1)$. The well is determined by the function $v|x^1 - x_1|^{-\rho}$ for all $x^1 \geq 0$ in the case 1^0 and for $x^1 \leq x_1 - 2^{1/2}c$ in the case 2^0; in the second case the behaviour of V for $x^1 \geq x_1 - 2^{1/2}c$ is inessential.

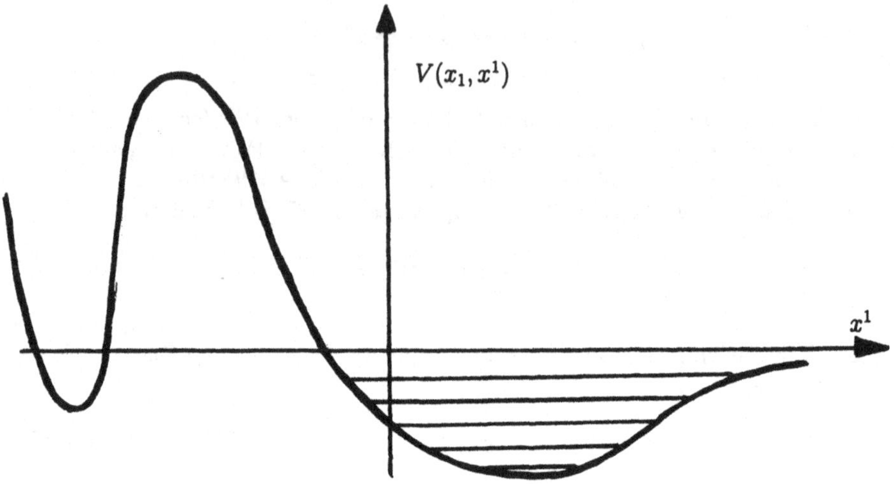

Figure 15.1. The case 1^0.

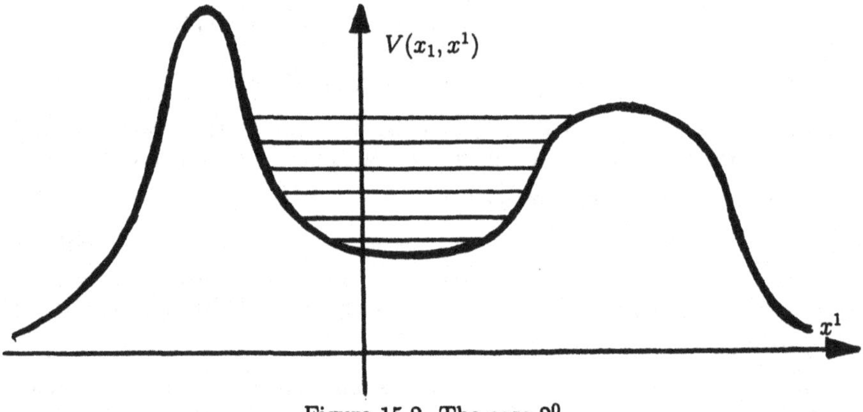

Figure 15.2. The case 2^0.

In both cases the potential well can be approximated by a linear function

$$v|x_1|^{-\rho} + |v|\rho|x_1|^{-\rho-1}x^1, \quad x^1 \geq 0.$$

After the same as in subsection 15.3 change of variables $x^1 \mapsto |x_1|^{\rho+1}x^1$, this leads to equation (15.17). The values of $\psi(x_1, x^1)$ for $x^1 \leq 0$ do not intervene into the final result but are essential for the proof. Actually, $\psi(x_1, x^1)$ decays as $\exp(-c|x^1|^r)$ for $x^1 \to -\infty$, and it is determined by quasi-classical formulas for all $x^1 < 0$ in the limit $|x_1| \to \infty$. Solutions of equation (15.2) for $x^1 < 0$ and $x^1 > 0$ are linked by the matching condition at the point $x^1 = 0$ which becomes $\psi(x_1, 0) = 0$ in the limit $x_1 \to -\infty$ in the case 1^0 and in the limit $x_1 \to \infty$ in the case 2^0.

Note that in the case 1^0 the potential (15.16) has also a well on the half-axis $x^1 \leq 0$ concentrated around the point x_1. This well gives rise to channels of scattering on bound states of the Hamiltonian H^2. Due to a positive barrier V^1, separating the wells on the positive and negative half-axes, the interaction between them can be neglected in the limit $x_1 \to -\infty$. So we can construct approximate eigenfunctions of the operator (15.1) concentrated on the half-axis $x^1 \geq 0$. The subspaces $\operatorname{Ran} w^\pm$ and $\operatorname{Ran} \mathcal{W}^\pm_{2,n}$ are orthogonal because the functions $JU_1(t)f_1$ tend to zero on compacts $|x^2| \leq C$ where the functions

$$\psi^{2,n}(x^2) \exp\big(i\Xi_{2,n}(x_2,t)\big)(2it)^{-1/2}\hat{f}_2(x_2/(2t))$$

are localized.

We emphasize that in the case 2^0 the operator $H(x_1)$ has no eigenvalues and $\psi(x_1)$ is generated by a positive minimum (in the variable x^1) of the potential $V(x_1, x^1)$. Thus $\psi(x_1)$ corresponds to a resonant state of the operator $H(x_1)$.

Note that for $f \in \operatorname{Ran} w^\pm$ the solution $u(t) = \exp(-iHt)f$ of the Schrödinger equation "lives" for large $|t|$ in the region where $x_1 \sim -|t|$ in the case 1^0 or $x_1 \sim |t|$ in the case 2^0 and in both cases $x^1 \sim |t|^\sigma$ for $\sigma \in (1/3, 1/2)$. Such solutions describe a physical process where a pair of particles (say, the first and the second) interacting by the potential V^1 are relatively close to one another and the third particle is far away. This pair is bound by a potential depending on the position of the third particle, but this bound state is evanescent as $|t| \to \infty$. Thus the asymptotic behaviour of solutions $u(t)$ for $f \in \operatorname{Ran} w^\pm$ resembles the behaviour of solutions for initial states $f \in \operatorname{Ran} \mathcal{W}^\pm_{1,n}$ (if the operator H^1 has bound states). However, similarly to the case $f \in \operatorname{Ran} \mathcal{W}^\pm_0$, in this channel the distances between all particles tend to infinity as $|t| \to \infty$.

One can imagine, for example, that the mass of the first particle is infinite, and it is fixed at the origin. Positions of the second and third particles with masses $1/2$ are given by coordinates x^1 and x_1, respectively. The second particle is retained on the positive half-axis by a potential barrier on the negative half-axis due to its interaction $V^1(x^1)$ with the first particle. A potential $V^2(2^{-1/2}(x^1 - x_1))$ of interaction between the second and third particles is long-range. In the case 1^0 (see Figure 15.3), V^2 is attactive and the third particle moves to minus infinity. In the case 2^0 (see Figure 15.4), V^2 is repulsive and the third particle moves to plus infinity. In both cases $|x_1|$ grows as $|t|$ for $|t| \to \infty$. The second particle is jammed to the first one by an interaction with the third particle. In the long run the second particle escapes to plus infinity but x^1 behaves as $|t|^\sigma$ for large $|t|$.

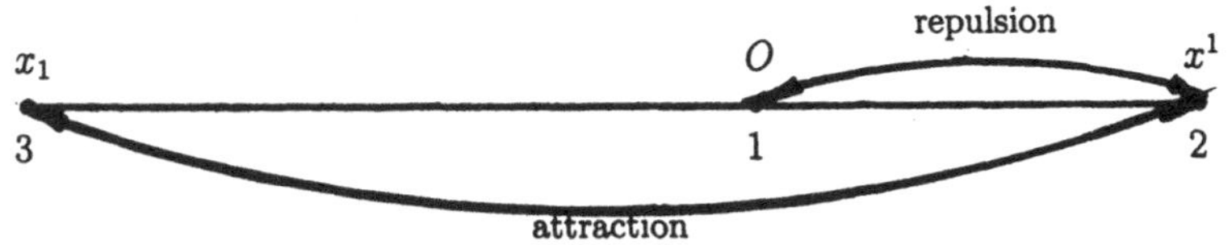

Figure 15.3. The case 1^0.

Figure 15.4. The case 2^0.

If the subspaces X^1 and X^2 are orthogonal, then due to the separation of variables the asymptotic completeness holds. In this case, our construction does not work because the potential V^2 in (15.18) does not depend on x^1.

We do not know whether the described quantum mechanical process has its counterpart in the classical mechanics.

Remark that the existence of new channels for three-particle systems automatically implies the same phenomena for systems of more than three particles. It suffices to take a system where all particles but three are free and the system of these three distinguished particles possesses a described channel.

As was pointed out already, the existence of wave operators $w^{\pm}$ contradicts the asymptotic completeness. It is less clear whether their existence is compatible with the asymptotic clustering (see subsection 11.4).

There is however a gap between the cases when asymptotic completeness holds and when it is violated. Hence the following questions arise.

Problem 15.4 *Is the scattering asymptotically complete when $\rho \in [1/2, \sqrt{3}-1)$? The same question for all $\rho < \sqrt{3}-1$ if particles are, say, three-dimensional.*

Note that, under some additional assumptions, asymptotic completeness for all $\rho > 1/2$ was checked in [150, 57].

In the cases when new channels are constructed one can expect that asymptotic completeness is recovered if these additional channels are taken into account. In a somewhat similar situation a result of such type was established in [156]. Thus, we formulate

Problem 15.5 *To prove (for example, under the assumptions of Theorems* 15.1, 15.2 *and* 15.3*) generalized asymptotic completeness, that is, that the ranges of all wave operators constructed exhaust $\mathcal{H}_H^{(ac)}$.*

16. THE HEISENBERG MODEL

The Heisenberg model of n interacting spins can be formulated in terms of a discrete version of the n-particle Schrödinger operator. The fundamental difference between multiparticle discrete and continuous Schrödinger operators is that in the discrete case the analogue of the operator $-\Delta$ is not rotationally invariant. The absence of this property impedes separation of variables (which was essential for considering non-interacting clusters of particles) and makes the formulation of asymptotic completeness more complicated. This is quite similar to dispersive Hamiltonians discussed in subsection 11.5. If $n = 2$, the problem can still be reduced to the one-particle case considered in subsection 2.4, but these new difficulties become quite substantial if $n > 2$.

16.1. The Hamiltonian. The Heisenberg model of a ferromagnet is described by a system of quantum mechanical spins, one at each point x of the lattice $\mathbf{Z}^k$ with the basis $e_1, \dots, e_k$. It is supposed that all spins point down, except some finite number of spins pointing up. Assume that we have exactly n "spins-up" called magnons, and let $x_1, x_2, \dots, x_n$ be the points where they are localized. The configuration space of n magnons consists of sets $\{x_1, x_2, \dots, x_n\}$, where all points $x_1, x_2, \dots, x_n \in \mathbf{Z}^k$ are different and their order is inessential. Quantum mechanical states are described by functions on the configuration space. Let us denote by D_n the subset of all points

$$\mathbf{x}_n = (x_1, x_2, \dots, x_n) \in \underbrace{\mathbf{Z}^k \times \mathbf{Z}^k \times \cdots \times \mathbf{Z}^k}_{n \text{ times}} =: \mathbf{Z}^{kn}$$

such that $x_i = x_j$ for at least one pair i, j, and let us put $B_n = \mathbf{Z}^{kn} \setminus D_n$. Thus the state space of n quantum mechanical magnons is the Hilbert space $L_2^{(s)}(B_n)$ of symmetric functions defined on the (symmmetric) set B_n.

The Heisenberg Hamiltonian $\mathbf{H}_n$ of n interacting magnons acts in the space $L_2^{(s)}(B_n)$. If $n = 1$, then $B_1 = \mathbf{Z}^k$ and

$$(\mathbf{H}_1 f)(x) = \sum_{|x'-x|=1} (f(x) - f(x')), \quad x, x' \in \mathbf{Z}^k,$$

so, up to the shift by $2kI$, it coincides with the discrete Laplacian. If $n \geq 2$, then

$$(\mathbf{H}_n f)(\mathbf{x}_n) = \sum_{|\mathbf{x}'_n - \mathbf{x}_n|=1, \mathbf{x}'_n \in B_n} (f(\mathbf{x}_n) - f(\mathbf{x}'_n)), \quad \mathbf{x}_n \in B_n. \tag{16.1}$$

Thus every magnon interacts only with its nearest neighbours. It is an exercice to check that the operator $2\mathbf{H}_n$ coincides with the restriction of the Heisenberg Hamiltonian defined in Chapter 14, v.3 of [134] on the n-magnon sector. The Hamiltonian

$\mathbf{H}_n^{(0)}$ of n non-interacting magnons acts in the space $L_2^{(s)}(\mathbf{Z}^{kn})$ of symmetric functions defined on the whole lattice $\mathbf{Z}^{kn}$ by the formula

$$(\mathbf{H}_n^{(0)} f)(\mathbf{x}_n) = \sum_{|\mathbf{x}'_n - \mathbf{x}_n| = 1} (f(\mathbf{x}_n) - f(\mathbf{x}'_n)), \quad \mathbf{x}_n \in \mathbf{Z}^{kn}. \tag{16.2}$$

We first neglect the symmetry and extend the operators $\mathbf{H}_n$ and $\mathbf{H}_n^{(0)}$ by the formulas (16.1) and (16.2) to the spaces $L_2(B_n)$ and $L_2(\mathbf{Z}^{kn})$, respectively. These operators will be denoted H_n and $H_n^{(0)}$. The subspaces $L_2^{(s)}(B_n) \subset L_2(B_n)$ and $L_2^{(s)}(\mathbf{Z}^{kn}) \subset L_2(\mathbf{Z}^{kn})$ of symmetric functions are invariant with respect to H_n and $H_n^{(0)}$, and

$$\mathbf{H}_n = H_n\big|_{L_2^{(s)}(B_n)}, \quad \mathbf{H}_n^{(0)} = H_n^{(0)}\big|_{L_2^{(s)}(\mathbf{Z}^{kn})}. \tag{16.3}$$

Let us introduce the shift in the variable x_j, $j = 1, \dots, n$:

$$(\mathrm{T}_j(p)f)(x_1, \dots, x_n) = f(x_1, \dots, x_{j-1}, x_j + p, x_{j+1}, \dots, x_n), \quad p \in \mathbf{Z}^k. \tag{16.4}$$

Then

$$H_n^{(0)} = -\sum_{j=1}^{n}\sum_{s=1}^{k}(\mathrm{T}_j(e_s) + \mathrm{T}_j(-e_s)) + 2knI. \tag{16.5}$$

It is sometimes convenient to extend the operator H_n by zero to the subspace $L_2(D_n) \subset L_2(\mathbf{Z}^{kn})$. The operator acting on the whole space $L_2(\mathbf{Z}^{kn})$ will be denoted $\hat{H}_n$. Formulas (16.1), (16.2) show that

$$\hat{H}_n = \chi_{B_n} H_n^{(0)} \chi_{B_n} + \mathcal{V}_n = H_n^{(0)} - \chi_{D_n} H_n^{(0)} - H_n^{(0)}\chi_{D_n} + \chi_{D_n} H_n^{(0)} \chi_{D_n} + \mathcal{V}_n, \tag{16.6}$$

where χ_{B_n}, χ_{D_n} are the operators of multiplication by the characteristic functions of the sets B_n, D_n, respectively, and $\mathcal{V}_n$ is multiplication by the function

$$\mathcal{V}_n(\mathbf{x}_n) = \begin{cases} -\sum_{|\mathbf{x}'_n - \mathbf{x}_n| = 1, \mathbf{x}'_n \in D_n} 1, & \mathbf{x}_n \in B_n, \\ 0, & \mathbf{x}_n \in D_n. \end{cases} \tag{16.7}$$

It follows from (16.5) and (16.6) that the operators $H_n^{(0)}$ and $\hat{H}_n$ are bounded and symmetric. Of course the same is true for all operators $\mathbf{H}_n^{(0)}$, H_n and $\mathbf{H}_n$.

Let the subset $D'_n \subset \mathbf{Z}^{kn}$ be defined by the condition: $\mathbf{x}_n \in D'_n$ if and only if $|x_i - x_j| \le 1$ for at least one pair i, j. By virtue of (16.6),

$$\hat{H}_n - H_n^{(0)} = (\hat{H}_n - H_n^{(0)})\chi_{D'_n}.$$

Set $d = kn$. Up to the shift by $2dI$, the operator $H_n^{(0)}$ coincides with operator (2.24) for $\mu_j = -1$. Therefore, as shown in subsection 2.4, it can be diagonalized by the discrete Fourier transform $\mathcal{F}$ defined by (2.26). Then it follows from (2.27) that

$$(\exp(-iH_n^{(0)}t)f)(x) = (2\pi)^{-d/2} e^{-2idt} \int_{\mathbf{T}^d} \exp(i\langle x, \xi\rangle + 2it\sum_{j=1}^{d}\cos\xi_j)(\mathcal{F}f)(\xi)d\xi.$$

Choosing functions $\mathcal{F}f$ from a suitable set of elements dense in $L_2(\mathbb{T}^d)$, we obtain (see Chapter 14, v.3 of [134], for details) that

$$||\chi_{D'_n} \exp(-iH_n^{(0)}t)f|| = O(|t|^{-m}), \quad \forall m. \tag{16.8}$$

Thus Cook's criterion (Proposition 1.6) gives us the following assertion.

Proposition 16.1 *For any n, the wave operators $W^{\pm}(\hat{H}_n, H_n^{(0)})$ exist.*

Let us denote $\mathcal{J}_n : L_2(\mathbf{Z}^{kn}) \to L_2(B_n)$ and $\mathcal{J}_n^{(s)} : L_2^{(s)}(\mathbf{Z}^{kn}) \to L_2^{(s)}(B_n)$ the operators of restriction on the set B_n. Since by (16.8)

$$s - \lim_{|t|\to\infty} (I_n - \mathcal{J}_n^* \mathcal{J}_n) \exp(-iH_n^{(0)}t) = 0,$$

Proposition 16.1 can be reformulated as an assertion about the existence and isometricity of wave operators $W^{\pm}(H_n, H_n^{(0)}; \mathcal{J}_n)$; moreover,

$$W^{\pm}(H_n, H_n^{(0)}; \mathcal{J}_n) = \mathcal{J}_n W^{\pm}(\hat{H}_n, H_n^{(0)}). \tag{16.9}$$

According to (16.3), this implies the existence and isometricity of wave operators

$$W^{\pm}(\mathbf{H}_n, \mathbf{H}_n^{(0)}; \mathcal{J}_n^{(s)}) = W^{\pm}(H_n, H_n^{(0)}; \mathcal{J}_n)\big|_{L_2^{(s)}(\mathbf{Z}^{kn})}. \tag{16.10}$$

For initial states from the range of wave operator (16.10), a system of n interacting magnons dissolves asymptotically into n magnons moving independently of each other. However, similarly to n particles described by the Schrödinger operator, the range of $W^{\pm}(\mathbf{H}_n, \mathbf{H}_n^{(0)}; \mathcal{J}_n^{(s)})$ might not coincide with the absolutely continuous subspace of $\mathbf{H}_n$. This means that other channels of scattering are possible.

16.2. Bound states. A more advanced analysis of the operator $\hat{H}_n$ requires consideration of "bound states" of spin systems. Note first of all that if all n magnons are displaced by the same vector $p \in \mathbf{Z}^k$, then properties of the system are not changed. Indeed, let $\mathbf{T}_n(p)$,

$$(\mathbf{T}_n(p)f)(x_1, x_2, \ldots, x_n) = f(x_1 + p, x_2 + p, \ldots, x_n + p),$$

(that is $\mathbf{T}_n(p) = \mathrm{T}_1(p) \cdots \mathrm{T}_n(p)$) be the operator of such translations. This operator acts in the space $L_2(\mathbf{Z}^{kn})$ but $L_2^{(s)}(\mathbf{Z}^{kn})$, $L_2(B_n)$ and $L_2^{(s)}(B_n)$ are its invariant subspaces. The operators $\hat{H}_n$ and $\mathbf{T}_n(p)$ commute:

$$\hat{H}_n \mathbf{T}_n(p) = \mathbf{T}_n(p) \hat{H}_n$$

for any $p \in \mathbf{Z}^k$. Therefore we can decompose the space $L_2(\mathbf{Z}^{kn})$ in a diagonal for the operators $H_n^{(0)}$ and $\hat{H}_n$ direct integral whose fibers are parametrized by $\xi \in \mathbb{T}^k = [0, 2\pi)^k$ and consist of functions on $\mathbf{Z}^{kn}$ satisfying the condition

$$(\mathbf{T}_n(p)f)(\mathbf{x}_n) = \exp(i\langle p, \xi\rangle) f(\mathbf{x}_n), \quad \forall p \in \mathbf{Z}^k.$$

The parameter ξ is naturally interpreted as the total momentum of n magnons.

More precisely, let us introduce a unitary mapping

$$F_n : L_2(\mathbf{Z}^{kn}) \to L_2(\mathbf{Z}^{k(n-1)}) \otimes L_2(\mathbf{T}^k) =: \tilde{\mathcal{H}}_n$$

by the equality

$$(F_n f)(y_1, \dots, y_{n-1}, \xi) = (2\pi)^{-k/2} \sum_{y \in \mathbf{Z}^k} f(y_1 + y, \dots, y_{n-1} + y, y) e^{-i\langle y, \xi \rangle}.$$

Then

$$(F_n^* g)(x_1, \dots, x_{n-1}, x_n) = (2\pi)^{-k/2} \int_{\xi \in \mathbf{T}^k} g(x_1 - x_n, \dots, x_{n-1} - x_n, \xi) e^{i\langle x_n, \xi \rangle} d\xi \quad (16.11)$$

and the operator $\tilde{H}_n = F_n \hat{H}_n F_n^*$ acts in the space $\tilde{\mathcal{H}}_n$ as multiplication by an operator-function $\hat{h}_n(\xi) : L_2(\mathbf{Z}^{k(n-1)}) \to L_2(\mathbf{Z}^{k(n-1)})$. Similarly, the operator $\tilde{H}_n^{(0)} = F_n H_n^{(0)} F_n^*$ acts as multiplication by $h_n^{(0)}(\xi)$ in the space $L_2(\mathbf{Z}^{k(n-1)})$.

This procedure is quite similar to the "separation" of center of mass motion for dispersive Hamiltonians in subsection 11.5. However, in the latter case the roles of $\mathbf{Z}^k$ and $\mathbf{T}^k$ were played by $\mathbf{R}^k$.

Let us calculate the operators $h_n^{(0)}(\xi)$ and $\hat{h}_n(\xi)$. According to (16.4), (16.11)

$$\mathrm{T}_j(p) F_n^* = F_n^* \mathrm{T}_j(p), \quad j = 1, \dots, n-1, \quad \mathrm{T}_n(p) F_n^* = e^{i\langle p, \xi \rangle} F_n^* \mathbf{T}_{n-1}(-p).$$

By virtue of (16.5), this implies that

$$\begin{aligned} h_n^{(0)}(\xi) = & - \sum_{j=1}^{n-1} \sum_{s=1}^{k} (\mathrm{T}_j(e_s) + \mathrm{T}_j(-e_s)) \\ & - \sum_{s=1}^{k} (e^{-i\langle e_s, \xi \rangle} \mathbf{T}_{n-1}(e_s) + e^{i\langle e_s, \xi \rangle} \mathbf{T}_{n-1}(-e_s)) + 2knI. \end{aligned} \quad (16.12)$$

Denote by $\tilde{D}_{n-1}$ the union of D_{n-1} and of the set of points

$$\mathbf{y}_{n-1} = (y_1, y_2, \dots, y_{n-1}) \in \mathbf{Z}^{k(n-1)}$$

such that $y_j = 0$ at least for one $j = 1, \dots, n-1$ and let $\chi_{\tilde{B}_{n-1}}$ be the characteristic function of the set $\tilde{B}_{n-1} = \mathbf{Z}^{k(n-1)} \setminus \tilde{D}_{n-1}$. Then

$$\chi_{B_n}(x_1, \dots, x_n) = \chi_{\tilde{B}_{n-1}}(x_1 - x_n, \dots, x_{n-1} - x_n),$$

and hence

$$(F_n \chi_{B_n} F_n^* g)(\mathbf{y}_{n-1}, \xi) = \chi_{\tilde{B}_{n-1}}(\mathbf{y}_{n-1}) g(\mathbf{y}_{n-1}, \xi).$$

It follows that the operator $F_n \chi_{B_n} H_n^{(0)} \chi_{B_n} F_n^*$ acts in the space $\tilde{\mathcal{H}}_n$ as multiplication by the operator-function

$$\chi_{\tilde{B}_{n-1}} h_n^{(0)}(\xi) \chi_{\tilde{B}_{n-1}} : L_2(\mathbf{Z}^{k(n-1)}) \to L_2(\mathbf{Z}^{k(n-1)}).$$

Clearly, the operator $\mathcal{V}_n$ of multiplication by function (16.7) commutes with $\mathbf{T}_n(p)$ for all $p \in \mathbb{Z}^k$. Therefore

$$\mathcal{V}_n(x_1, \ldots, x_n) = v_n(x_1 - x_n, \ldots, x_{n-1} - x_n),$$

where

$$v_n(y_1, \ldots, y_{n-1}) = \mathcal{V}_n(y_1, \ldots, y_{n-1}, 0). \tag{16.13}$$

In particular, $v_n(\mathbf{y}_{n-1}) = 0$ if $\mathbf{y}_{n-1} \in \tilde{D}_{n-1}$. Consequently, the operator $F_n \mathcal{V}_n F_n^*$ acts as multiplication by the function $v_n(\mathbf{y}_{n-1})$ (for all $\xi \in \mathbb{T}^k$). Now formula (16.6) allows us to put together the results obtained.

Proposition 16.2 *Let the operators $h_n^{(0)}(\xi)$ and v_n in the space $L_2(\mathbb{Z}^{k(n-1)})$ be defined by formulas* (16.12) *and* (16.13), *respectively. Then the operators $\tilde{H}_n^{(0)} = F_n H_n^{(0)} F_n^*$ and $\tilde{H}_n = F_n \hat{H}_n F_n^*$ act in the space $\tilde{\mathcal{H}}_n$ as multiplication by the operator-functions $h_n^{(0)}(\xi)$ and*

$$\hat{h}_n(\xi) = \chi_{\tilde{B}_{n-1}} h_n^{(0)}(\xi) \chi_{\tilde{B}_{n-1}} + v_n. \tag{16.14}$$

Note that the operator F_n can be considered as a unitary mapping of the space $L_2(B_n)$ onto $L_2(\tilde{B}_{n-1}) \otimes L_2(\mathbb{T}^k)$. It follows from Proposition 16.2 that the operator $F_n H_n F_n^*$ acts in this space as multiplication by the operator-function (16.14) which will be denoted $h_n(\xi)$ in this case; so $h_n(\xi) : L_2(\tilde{B}_{n-1}) \to L_2(\tilde{B}_{n-1})$.

The point spectrum of the operators $h_n(\xi)$ gives rise, normally, to the absolutely continuous spectrum of the operator H_n. Indeed, suppose that operators $h_n(\xi)$, for all ξ from some subset $U_n^{(j)} \subset \mathbb{T}^k$, have eigenvalues $\lambda_n^{(j)}(\xi)$, $j = 1, 2, \ldots$, with corresponding normalized eigenfunctions $\psi_n^{(j)}(\xi)$. Let $\tilde{\mathcal{H}}_n^{(j)}$ be spanned by functions $\psi_n^{(j)}(\mathbf{y}_{n-1}, \xi) f_n^{(j)}(\xi)$ where $f_n^{(j)}$ is an arbitrary function from $L_2(U_n^{(j)})$ and

$$\tilde{\mathcal{H}}_n^{(bs)} = \bigoplus_j \tilde{\mathcal{H}}_n^{(j)}.$$

Every subspace $\tilde{\mathcal{H}}_n^{(j)}$ is isomorphic to $L_2(U_n^{(j)})$, and the operator $F_n H_n F_n^*$ reduces to multiplication by $\lambda_n^{(j)}(\xi)$ on it. Let us denote by $H_n^{(bs)}$ the restriction of the operator H_n on the subspace $\mathcal{H}_n^{(bs)} = F_n^* \tilde{\mathcal{H}}_n^{(bs)}$. We expect that the functions $\lambda_n^{(j)}(\xi)$ cannot take constant values on sets of positive measure. This can be formulated (cf. subsection 11.5) as

Problem 16.3 *Prove that, for any n, the operator $H_n^{(bs)}$ is absolutely continuous.*

This problem seems to be similar in spirit to the absence of eigenvalues of the Schrödinger operator with a periodic potential (see [148] and v.4 of [134]).

The vectors from the subspaces $\mathcal{H}_n^{(bs)}$ and

$$\mathcal{H}_n^{(scat)} = L_2(B_n) \ominus \mathcal{H}_n^{(bs)}$$

play, respectively, the roles of bound and scattering states for the multiparticle Schrödinger operator. To take the symmetry into account, we introduce the orthogonal projection $P_n^{(s)}$ of the space $L_2(B_n)$ on its symmetric subspace $L_2^{(s)}(B_n)$. Then the vectors from the subspaces $P_n^{(s)}\mathcal{H}_n^{(bs)}$ and $P_n^{(s)}\mathcal{H}_n^{(scat)}$ are, respectively, bound and scattering states of n magnons described by the Hamiltonian $\mathbf{H}_n$.

16.3. Two interacting magnons. Proposition 16.2 reduces the Hamiltonian $\hat{H}_2$ of two interacting magnons to the family $\hat{h}_n(\xi)$, $\xi \in \mathbf{T}^k$, of one-particle discrete Schrödinger operators in the space $L_2(\mathbf{Z}^k)$. Consider first the "free" Hamiltonian $H_2^{(0)}$. According to (16.12) the operator $F_2 H_2^{(0)} F_2^*$ acts as multiplication by the operator-function

$$h_2^{(0)}(\xi) = -\sum_{s=1}^{k}(\mu_s(\xi)\mathrm{T}_j(e_s) + \bar{\mu}_s(\xi)\mathrm{T}_j(-e_s)) + 4kI, \quad \mu_s(\xi) = 1 + e^{-i\langle e_s, \xi\rangle},$$

in the space $\tilde{\mathcal{H}}_2 = L_2(\mathbf{Z}^k) \otimes L_2(\mathbf{T}^k)$.

The function $v_2(y)$ (see (16.7), (16.13)) equals 0 for $|y| \neq 1$ and it equals -2 for $|y| = 1$. The set $\tilde{D}_1$ consists of the single point $y = 0$. Consequently, it follows from Proposition 16.2 that

$$\hat{h}_2(\xi) = h_2^{(0)}(\xi) + \mathbf{v}_2(\xi)$$

where

$$(\mathbf{v}_2(\xi)g)(y) = -(h_2^{(0)}(\xi)g)(0)\chi_{\tilde{D}_1}(y) - g(0)(h_2^{(0)}(\xi)\chi_{\tilde{D}_1})(y) + \tilde{v}_2(y)g(y),$$

$\tilde{v}_2(0) = 4k$, $\tilde{v}_2(y) = -2$ for $|y| = 1$ and $\tilde{v}_2(y) = 0$ for $|y| > 1$. Thus $\mathbf{v}_2(\xi)$ is a finite-dimensional operator (of rank $2k+1$), and it satisfies the assumptions of Theorem 2.10. Both Theorems 1.7 and 2.10 imply that the wave operators $W^{\pm}(\hat{h}_2(\xi), h_2^{(0)}(\xi))$ exist for all $\xi \in \mathbf{T}^k$ and are complete. Moreover, the second of these theorems shows that the singular continuous part of the spectrum of the operator $\hat{h}_2(\xi)$ is empty and its point spectrum may accumulate at the thresholds of $h_2^{(0)}(\xi)$ only. Since the operator

$$F_2 \exp(\hat{H}_2 t)\exp(-H_2^{(0)} t) F_2^*$$

acts in the space $\tilde{\mathcal{H}}_2$ as multiplication by the operator-function

$$\exp(\hat{h}_2(\xi)t)\exp(-h_2^{(0)}(\xi)t) : L_2(\mathbf{Z}^k) \to L_2(\mathbf{Z}^k),$$

the wave operators $W^{\pm}(\hat{H}_2, H_2^{(0)})$ exist. Furthermore, $F_2 W^{\pm}(\hat{H}_2, H_2^{(0)}) F_2^*$ act in the space $\tilde{\mathcal{H}}_2$ as multiplication by $W^{\pm}(\hat{h}_2(\xi), h_2^{(0)}(\xi))$. According to (16.9), this leads to

Theorem 16.4 *The wave operators $W^{\pm}(H_2, H_2^{(0)}; \mathcal{J}_2)$ exist and $F_2 W^{\pm}(H_2, H_2^{(0)}; \mathcal{J}_2)$ F_2^* act in the space $\tilde{\mathcal{H}}_2$ as multiplication by $W^{\pm}(h_2(\xi), h_2^{(0)}(\xi); \tilde{J}_2)$, where $\tilde{J}_2 : L_2(\mathbf{Z}^k) \to L_2(\tilde{B}_1)$ is the operator of restriction on the set $\tilde{B}_1 = \mathbf{Z}^k \setminus \{0\}$. In particular, the operators $W^{\pm}(H_2, H_2^{(0)}; \mathcal{J}_2)$ are isometric, and their ranges*

$$\operatorname{Ran} W^{\pm}(H_2, H_2^{(0)}; \mathcal{J}_2) = \mathcal{H}_2^{(scat)}.$$

This result was first obtained in [61]. It follows from Theorem 16.4 that the restriction of the operator H_2 on the subspace $\mathcal{H}_2^{(scat)}$ is absolutely continuous. We recall that, on its orthogonal complement $\mathcal{H}_2^{(bs)}$, the operator $H_2^{(bs)} = H_2\big|_{\mathcal{H}_2^{(bs)}}$ is unitarily equivalent to the orthogonal sum of operators of multiplication by functions $\lambda_2^{(j)}(\xi)$, $j = 1, 2, \ldots$. We emphasize that Problem 16.3 seems to be open even for $n = 2$. By virtue of (16.3), (16.10), all the results above on the operator H_2 remain valid for the operator $\mathbf{H}_2$.

16.4. Asymptotic completeness. The scattering problem for n interacting magnons is not solved although its plausible formulation seems to be clear. Let us first neglect symmetry and consider the operator $H = H_n$ acting in the space $L_2(B_n)$. Let a be a decomposition of n magnons into noninteracting clusters C_l, that is

$$C_1 \cup \ldots \cup C_p = \{1, \ldots, n\}, \quad C_i \cap C_j = \emptyset \quad \text{if} \quad i \neq j, \quad p > 1.$$

Let $\kappa_l = |C_l|$ be the number of elements in the set C_l. We associate to every $C_l = \{i_1, \ldots, i_{\kappa_l}\}$ the space $\mathcal{H}_{C_l}$ and the Hamiltonian H_{C_l} acting in this space. The space $\mathcal{H}_{C_l}$ consists of functions of the variables $x_{i_1}, \ldots, x_{i_{\kappa_l}}$. It is isomorphic to $L_2(\mathbf{Z}^k)$ if $\kappa_l = 1$ and isomorphic to $\mathcal{H}_{\kappa_l}^{(bs)}$ if $\kappa_l > 1$. The Hamiltonian H_{C_l} equals $H_1^{(0)}$ if $\kappa_l = 1$ and equals $H_{\kappa_l}^{(bs)}$ if $\kappa_l > 1$. Now we define

$$\mathcal{H}_a = \bigotimes_{l=1}^{p} \mathcal{H}_{C_l}$$

and

$$H_a = H_{C_1} \otimes I \otimes \ldots \otimes I + I \otimes H_{C_2} \otimes I \otimes \ldots \otimes I + \ldots + I \otimes \ldots \otimes I \otimes H_{C_p}.$$

The space $\mathcal{H}_a$ can of course be viewed as a subspace of $L_2(\mathbf{Z}^{kn})$. We denote by $J_a : \mathcal{H}_a \to L_2(B_n)$ the restriction of the operator $\mathcal{J}_n$ on this space, that is $(J_a f)(\mathbf{x}_n) = f(\mathbf{x}_n)$ if $\mathbf{x}_n \in B_n$ and $(J_a f)(\mathbf{x}_n) = 0$ if $\mathbf{x}_n \in D_n$.

We conjecture that all wave operators $W^{\pm}(H, H_a; J_a)$ exist, are isometric, their ranges are orthogonal and the asymptotic completeness

$$\bigoplus_a \operatorname{Ran} W^{\pm}(H, H_a; J_a) = \mathcal{H}_n^{(scat)} \tag{16.15}$$

holds. The sum here is taken over all non-trivial decompositions a of the set $\{1, \ldots, n\}$.

Equality (16.15) implies that the restriction of the operator H_n on the subspace $\mathcal{H}_n^{(scat)}$ is absolutely continuous. Therefore we can reformulate Problem 16.3 in a stronger form.

Problem 16.5 *Prove that, for any n, the operator H_n is absolutely continuous.*

We emphasize that, in the case $\kappa_l = 1$, the Hamiltonian H_{C_l} describes a free magnon and, in the case $\kappa_l > 1$, it describes a system of κ_l magnons forming a bound state between themselves but moving with some common momentum. Thus such states should belong to the absolutely continuous subspace of the operator H_{C_l}. The Hamiltonian H_a describes a system of such clusters with interactions between different clusters neglected. In particular, if $p = n$, then $H_a = H_n^{(0)}$. Note also that $H_a = H_n^{(bs)}$ for $p = 1$. Physically, the asymptotic completeness means that a system of n interacting magnons may either form a bound state or it splits into non-interacting subsystems of magnons forming a bound state each and of some number of free magnons.

The formulation of the asymptotic completeness for the operator $\mathbf{H}_n$ differs from (16.15) only by restriction of all operators on the subspaces of symmetric functions (cf. (16.10)). For example, if $n = 4$, then there are 4 possibilities:

1^0 $p = 4$ (all magnons are asymptotically free)

2^0 $p = 3$ (two magnons are free and two others are in a bound state)

3^0 $p = 2$, $\kappa_1 = 1$, $\kappa_2 = 3$ (one magnon is free and three others form a bound state)

4^0 $p = 2$, $\kappa_1 = \kappa_2 = 2$ (there are two pairs of two magnons in a bound state each, and these pairs move independently of each other).

It is also possible that all four magnons form a bound state, but this state belogs to the subspace $\mathcal{H}_4^{(bs)}$.

Thus we finish with

Problem 16.6 *Prove the existence of the wave operators* $W^{\pm}(H, H_a; J_a)$ *and the asymptotic completeness* (16.15) *in the Heisenberg model.*

17. INFINITE OBSTACLE SCATTERING

Scattering on a bounded obstacle can be treated essentially by the same methods as the corresponding problem for the Schrödinger operator with a short-range (even of compact support) potential. For example, trace-class methods of [10] suit perfectly to that purpose; see the books [134], v.3, and [36] for a comprehensive presentation of this theory. We emphasize that for bounded obstacles a type of boundary conditions is inessential.

On the contrary, scattering on unbounded obstacles acquires some features of the multiparticle problem. An important point in the proof of asymptotic completeness is a verification that surface waves do not exist. In the general case (except cones) only the Dirichlet boundary condition was considered [32, 74, 75].

17.1. Formulation of the problem. Let Ω be a domain with (noncompact) boundary $\partial\Omega$ of class C^2. It is supposed that Ω satisfies the illumination condition, that is $\langle x, \nu(x)\rangle \leq 0$ for the unit vector $\nu(x)$ of the outer normal if $x \in \partial\Omega$ and $|x|$ is sufficiently large ($|x| \geq r_0$). Geometrically this condition means that a ray passing from the origin to the point x remains on its continuation in $\bar{\Omega}$. Set

$$G_r = \{\omega \in \mathbf{S}^{d-1} : x = r\omega \in \Omega\} \quad \text{and} \quad G = \bigcup_{r \geq r_0} G_r.$$

The cone $K = K_\Omega = \{x \in \mathbf{R}^d \setminus \{0\} : x = r\omega, \omega \in G\}$ is called the limit cone for Ω. For example, one can keep in mind the case when the obstacle $\mathcal{O} = \mathbf{R}^d \setminus \Omega$ is a paraboloid. Then G consists of the whole sphere $\mathbf{S}^{d-1}$ with only one point removed; see Figure 17.1 where $\mathcal{O}$ is shadowed.

Our goal is to compare the operator $H = -\Delta$ with the Dirichlet boundary condition $u(x) = 0$ for $x \in \partial\Omega$ in the Hilbert space $\mathcal{H} = L_2(\Omega)$ with the unperturbed operator $H_0 = -\Delta$ acting in the space $\mathcal{H}_0 = L_2(\mathbf{R}^d)$. Since the operators H_0 and H act in different spaces, a formulation of the scattering problem requires an identification $J : \mathcal{H}_0 \to \mathcal{H}$. It is natural to assume that J acts as the restriction of a function $f \in L_2(\mathbf{R}^d)$ on the set Ω. Then $J^* : \mathcal{H} \to \mathcal{H}_0$ extends a function $f \in L_2(\Omega)$ by zero to all $\mathbf{R}^d$. The results of [32] and [74, 75] are formulated in the following

Theorem 17.1 *The wave operators $W^\pm = W^\pm(H, H_0; J)$ exist, are complete, that is* $\operatorname{Ran} W^\pm = \mathcal{H}$ *(the operator H is absolutely continuous), and*

$$||W^\pm f|| = ||\chi_\pm \hat{f}||, \tag{17.1}$$

where $\chi_\pm(\xi) = \chi_K(\pm\xi)$, χ_K is the characteristic function of K and $\hat{f}$ is the Fourier transform of f. In particular, the operators $W^\pm$ are isometric if the measure $|\mathbf{S}^{d-1} \setminus G| = 0$. The wave operators $W^\pm(H_0, H; J^*)$ also exist and are isometric.

Similarly to potential scattering, a proof of the existence of direct wave operators $W^{\pm}$ as well as of equality (17.1) can easily be obtained by the Cook method. In particular, the method of proof of Theorem 3.10 can conveniently be adjusted to the obstacle scattering. On the other hand, the completeness of $W^{\pm}$ or the existence and isometricity of the "inverse" wave operators $W^{\pm}(H_0, H; J^*)$ poses a substantial problem. Methods of [32] and [74, 75] of its solution are rather different although both of them use the limiting absorption principle and radiation estimates. The paper [32] relies on the construction of the generalized Fourier transform (cf. Section 5) and [74, 75] fits the obstacle scattering into the framework of smooth perturbations theory (cf. Section 4).

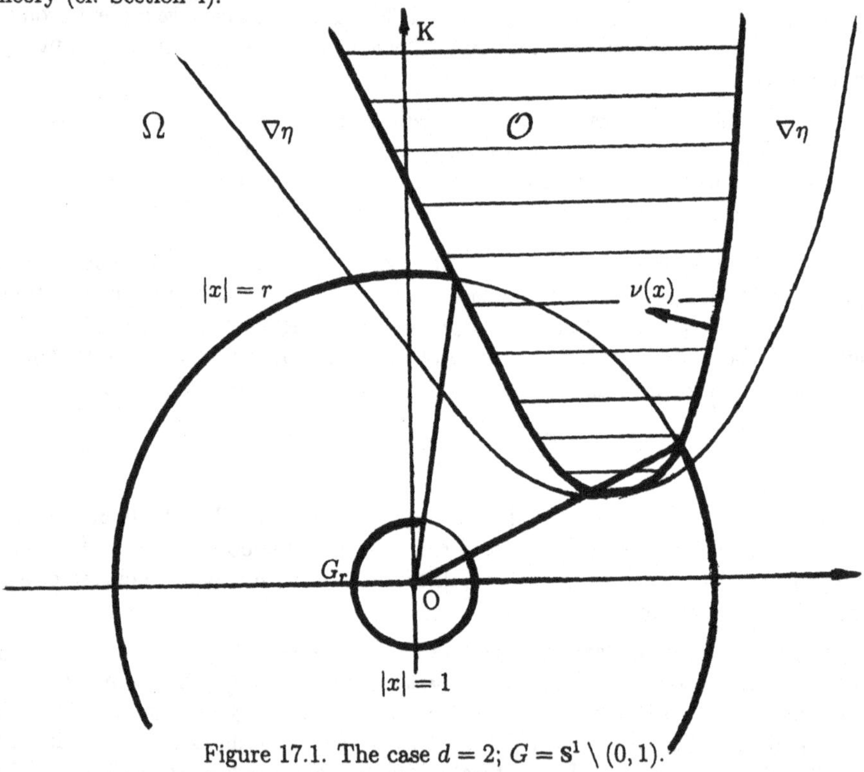

Figure 17.1. The case $d = 2$; $G = \mathbb{S}^1 \setminus (0,1)$.

17.2. The scheme of smooth perturbations. The initial point of [74, 75] is the following analytical result which extends earlier results of [43, 151] (see also [152]). Its proof requires some mild additional technical assumptions compared to the illumination condition. In particular, it is supposed that

$$\operatorname{dist}(x, \partial K) = O(|x|^{\gamma}), \quad x \in \partial\Omega, \quad \gamma \in (0,1). \tag{17.2}$$

Proposition 17.2 *Let $z = k^2 \pm i\varepsilon$ where $\varepsilon \in (0,1]$, $k \in [k_0, k_1]$, $0 < k_0 < k_1 < \infty$ and $\alpha \in (-1/2, 1/2)$. Then the following inequalities hold*

$$\|\langle x\rangle^{\alpha}(\nabla \mp ik\hat{x})R(z)f\| \le C\|\langle x\rangle^{\alpha+1}f\|, \tag{17.3}$$

$$\|\langle x\rangle^{\alpha+1/2}(\nabla \mp ik\hat{x})R(z)f\| \leq C\varepsilon^{-1/2}\|\langle x\rangle^{\alpha+1}f\|. \tag{17.4}$$

Moreover, for a suitable set of elements f,

$$\lim_{\varepsilon\to 0}\varepsilon^{1/2}\|\langle x\rangle^{\alpha+1/2}(\nabla \mp ik\hat{x})R(k^2 \pm i\varepsilon)f\| = 0 \tag{17.5}$$

and the function $R(z)f$ has a limit as $\varepsilon \to 0$ in H^1_{loc} uniformly in $k \in [k_0, k_1]$. In particular, the spectrum of the operator H is absolutely continuous.

By the proof of Theorem 17.1, Proposition 17.2 plays the same role as Theorems 4.2 and 4.3 for construction of scattering theory for long-range potentials (and Theorems 13.1 and 13.2 for construction of scattering theory in the multiparticle case). However, the formulation of radition conditions in Proposition 17.2 is somewhat different from that given by Theorems 4.3 (and 9.1). Indeed, Theorems 4.3 and 9.1 correspond to the limit case $\alpha = -1/2$ of estimate (17.3). If $f \in \mathcal{S}$, then the estimate of $\langle x\rangle^\alpha(\nabla \mp ik\hat{x})R(z)f$ is getting more difficult as α increases. The condition $\alpha < 1/2$ is optimal since for $\alpha = 1/2$ estimate (17.3) is violated even for the resolvent of the free operator H_0. On the other hand, the only case $\alpha = -1/2$ which was needed for long-range and multiparticle potentials is excluded in Proposition 17.2. Furthermore, Proposition 17.2 contains an information on the function $(\partial_r \mp ik)R(z)f$ while Theorems 4.3 and 9.1 are formulated in terms of $\nabla^\perp R(z)f$ only. The proof of Proposition 17.2 relies on the integration-by-parts machinery which is resemblent but, strictly speaking, different from the commutator method used in Sections 4 and 13.

For the proof of the existence of $W^\pm(H_0, H; J^*)$, E. M. Il'in first considers the wave operators $W^\pm(H_0, H; J_1)$ where J_1 is multiplication by a C^∞-function η, $0 \leq \eta(x) \leq 1$, which equals to zero in a neighbourhood of $\partial\Omega$ and $\eta(x) = 1$ off a "parabolic" neighbourhood $\Omega_{\delta,c} \subset \Omega$ of $\partial\Omega$ of order $\delta \in [\gamma, 1)$ determined by the condition $\mathrm{dist}(x, \partial\Omega) \leq c|x|^\delta$. Then

$$\partial_j\eta(x) = O(|x|^{-\delta}), \quad \partial_j\partial_k\eta(x) = O(|x|^{-2\delta}), \quad \partial\eta(x)/\partial r = O(|x|^{-1}). \tag{17.6}$$

For example, if $\partial\Omega$ is given by the equation $x_d = c(x_1^2 + \ldots + x_{d-1}^2)^{\beta/2}$, $\beta > 1$, then $K = \partial K$ is the half-line $x_1 = \ldots = x_{d-1} = 0$, $x_d \geq 0$, condition (17.2) is satisfied for $\gamma = \beta^{-1}$ and η can be defined by the formula $\eta(x) = \theta(x_d(x_1^2 + \ldots + x_{d-1}^2)^{-\delta/2})$ where $\theta(s) = 1$ for small $|s|$ and $\theta(s) = 0$ for large $|s|$.

Recall (see Proposition 2.1) that the operator $\langle x\rangle^{-l}$ for any $l > 1/2$ is H_0-smooth on any bounded disjoint from zero interval. Therefore a minor modification of Proposition 1.18 shows that the wave operators $W^\pm(H_0, H; J_1)$ exist provided

$$\int_{-\infty}^{\infty}\|\langle x\rangle^l T_1\exp(-iHt)f\|^2 dt < \infty, \quad T_1 = H_0J_1 - J_1H.$$

for some set of elements f dense in $\mathcal{H}$ and some $l > 1/2$. It is not difficult to show (cf. different definitions of H-smoothness) that the last condition is satisfied if

$$\sup_{\pm\mathrm{Im}\, z>0, \mathrm{Re}\, z\in[k_0,k_1]}\|\langle x\rangle^l T_1 R(z)f\| < \infty \tag{17.7}$$

for any $0 < k_0 < k_1 < \infty$. To verify this condition, let us calculate

$$T_1 = -2(\nabla\eta)\nabla - (\Delta\eta) = -2(\nabla\eta)(\nabla \mp ik\hat{x}) \mp 2ik(\partial_r\eta) - (\Delta\eta) \qquad (17.8)$$

and use (17.6). Estimate (17.3) shows that the contribution of the first term in the right-hand side of (17.8) to (17.7) is bounded if $l < 1/2 + \delta$. To estimate the contributions of the two other terms in (17.8) we should additionally take into account that supports of $\partial_r\eta$ and $\Delta\eta$ belong to the parabolic region $\Omega_{\delta,c}$. In this region an inequality

$$\int_{\Omega_{\delta,c}} |x|^{2(\alpha-\delta)}|v(x)|^2 dx \le C \int_\Omega |x|^{2\alpha}|\nabla v(x)|^2 dx, \quad \alpha < 1/2, \qquad (17.9)$$

of the Hardy type holds. We apply it to the function

$$v(x) = e^{\mp ik|x|}(R(z)f)(x)$$

and $\alpha = l - 1 + \delta$ (for the term containing $\partial_r\eta$) or $\alpha = l - \delta$ (for the term containing $\Delta\eta$). This gives (17.7) for $l < \min\{3/2 - \delta, 1/2 + \delta\}$.

To prove that the wave operators $W^\pm(H_0, H; J_1)$ are isometric, we shall verify that

$$\lim_{|t|\to\infty} (N \exp(-iHt)g, \exp(-iHt)g) = 0, \quad N = I - J_1^* J_1, \qquad (17.10)$$

for functions g from some set dense in $\mathcal{H}$. Since the existence of $W^\pm(H_0, H; J_1)$ implies the existence of the limit in (17.10), it suffices to check that the limit in the sense of Abel is equal to zero. It follows from the spectral theorem and the Parseval equality that for any $\varepsilon > 0$

$$2\varepsilon \int_0^\infty e^{-2\varepsilon t}(N \exp(\mp iHt)g, \exp(\mp iHt)g)dt = \pi^{-1}\varepsilon \int_0^\infty (NR(\lambda \pm i\varepsilon)g, R(\lambda \pm i\varepsilon)g)d\lambda$$

$$\le (\pi^{-1}\varepsilon \int_0^\infty ||NR(\lambda \pm i\varepsilon)g||^2 d\lambda)^{1/2}||g||. \qquad (17.11)$$

Let us apply this inequality to an element $g = E(\Lambda)f$ where Λ is a bounded disjoint from zero interval and f belongs to the dense set where (17.5) is satisfied. Then the right-hand side of (17.11) tends to zero as $\varepsilon \to 0$ by virtue of an inequality of type (17.9). Thus (17.10) is true.

Relation (17.10) ensures that

$$\lim_{|t|\to\infty} \int_{\Omega_{\delta,c}} |(\exp(-iHt)f)(x)|^2 dx = 0, \quad \forall \delta \in [\gamma, 1), \qquad (17.12)$$

and hence

$$W^\pm(H_0, H; J^*) = W^\pm(H_0, H; J_1).$$

The meaning of (17.12) is that all states quit the preboundary zone, that is no surface waves exist which a priori could have violated the asymptotic completeness.

17.3. The generalized Fourier transform. The paper [32] follows the scheme developed earlier by W. Jäger [85, 86, 87] for the case of compact $\mathcal{O}$. The analytical background of this paper is almost the same as in [74, 75]. One of technical differences is that the limiting absorption principle and radiation estimates are formulated in [32] in terms of the spaces $\mathbf{B}$, $\mathbf{B}^*$ and $\mathbf{B}_0^*$ introduced for the case $\Omega = \mathbb{R}^d$ in subsection 3.4. These definitions extend naturally to the general case. The limiting absorption principle is formulated as inequality (3.11); moreover, the vector-function $\langle x\rangle^{-l}R(z)f$ for any $f \in \mathbf{B}$ and $l > 1/2$ is strongly continuous in z in the closed complex plane cut along $[0,\infty)$ with exception of the point $z = 0$. The radiation condition is formulated as the inclusion

$$(\partial_r \mp ik)R(k^2 \pm i0)f \in \mathbf{B}_0^*, \quad \forall f \in \mathbf{B}.$$

Similarly to the construction of Section 5, to define the generalized Fourier transform, P. Constantin proceeds from the asymptotics of the function $(R(k^2 \pm i0)f)(x)$ (it is extended by zero to the whole $\mathbb{R}^d$) as $|x| \to \infty$. Actually, he shows that

$$(R(k^2 \pm i0)f)(x) = \pi^{1/2}k^{-1/2}a^{\pm}(\pm\hat{x})|x|^{-(d-1)/2}e^{\pm ik|x| \mp i(d-3)4} + \varepsilon(x),$$

where $a \in L_2(G)$ and $\varepsilon \in \mathbf{B}_0^*$ (cf. Theorem 5.1). Let us set now $\mathfrak{N}_{\pm} = L_2(\pm G)$, $\mathfrak{H}_{\pm} = L_2(\mathbb{R}_+; \mathfrak{N}_{\pm})$ and define the operators $\Gamma^{\pm}(\lambda)$ and $F^{\pm} : \mathcal{H} \to \mathfrak{H}_{\pm}$ by equations (5.6) and (2.14). Then the equalities (5.7), (2.16), (2.17) and (2.19) are preserved (cf. Proposition 5.3 and Theorem 5.6). In the case $\Omega = \mathbb{R}^d$ the operators $F^{\pm}$ reduce of course to the operator F_0 defined in subsection 2.3 (that is, up to a change of variables, to the classical Fourier transform).

A next step in the approach of [32] is calculation of wave operators $W^{\pm}(H, H_0; J_0)$ where J_0 is multiplication by a C^∞-function η such that $\operatorname{supp}\eta \subset \Omega$, $\eta(x) = \theta(\hat{x})$ for large $|x|$ and $\theta \in C_0^\infty(G)$. It is shown that

$$W^{\pm}(H, H_0; J_0) = (F^{\pm})^*\theta^{\pm}F_0, \tag{17.13}$$

where $\theta^{\pm}$ is multiplication by the function $\theta(\pm\hat{\xi})$. Finally, approximating the characteristic function χ_G by functions $\theta_n \in C_0^\infty(G)$ P. Constantin shows that

$$W^{\pm}(H, H_0; J) = (F^{\pm})^*\chi^{\pm}F_0, \tag{17.14}$$

where $\chi^{\pm}$ is multiplication by $\chi_G(\pm\hat{\xi})$. Equations (17.13) and (17.14) play the role of (5.26) in potential scattering. By virtue of equality (2.16) the wave operators are complete which implies the existence of $W^{\pm}(H_0, H; J^*)$.

We emphasize that the illumination condition is the only assumption of [32] on the domain Ω. It was required there that $d \geq 3$, but this restriction is probably not necessary.

17.4. An arbitrary boundary condition. The papers [32] and especially [74, 75] treat fairly general differential operators. Moreover, the invariance principle (see equality (1.26)), proved for obstacle scattering in [74, 75], implies results on the wave equation.

The assumption that H is determined by the Dirichlet boundary condition plays an important role in these papers. Therefore the following question naturally arises.

Problem 17.3 *Develop scattering theory for the operator $H = -\Delta$ in a domain Ω with noncompact boundary $\partial\Omega$ (for example, for the exterior of a paraboloid) for the Neumann or more general boundary conditions*

$$\partial u/\partial\nu(x) + h(x)u = 0, \quad x \in \partial\Omega, \quad h = \bar{h}. \tag{17.15}$$

Of course, the wave operators $W^{\pm}(H, H_0; J)$ exist and equality (17.1) holds for an arbitrary function h. On the contrary, results on their completeness are scarce. The case where Ω is a cone was considered in [75]. It was verified there that conclusions of Theorem 17.1 hold if the function h in (17.15) satisfies the estimates

$$h(x) = O(|x|^{-\rho}), \quad \partial h/\partial r = O(|x|^{-\rho-1}), \quad \rho > 3/4,$$

as $|x| \to \infty$. Possibly, the method of [75] works for arbitrary $\rho > 1/2$.

On the other hand, one cannot expect that the wave operators $W^{\pm}(H, H_0; J)$ are complete without any assumptions on the function h. Indeed, suppose that Ω is a cone and $h(x) = h_0|x|^{-\rho}$ for some $\rho < 1/2$ and $h_0 < 0$ (and $|x|$ sufficiently large). Then all assumptions of subsection 15.1 are satisfied. Hence additional wave operators (15.6) exist and their ranges are orthogonal to those of $W^{\pm}(H, H_0; J)$ which contradicts the completeness of $W^{\pm}(H, H_0; J)$. We emphasize that in this case the equation $-\Delta u = \lambda u$ has solutions satisfying (17.15) and living in a parabolic neighbourhood of $\partial\Omega$. On the other hand, if $h(x) \geq 0$ or $h(x) = O(|x|^{-\rho})$ for some $\rho > 1/2$, then the construction of subsection 15.1 does not work, and one can expect that the wave operators $W^{\pm}(H, H_0; J)$ are complete.

We note, finally, that in [75] conditional theorems were established where estimates (17.3) and (17.4) were taken for assumptions (and the Dirichlet boundary condition was not explicitly required). In particular, the scheme of this paper can automatically be applied to the case (17.15) *provided* the estimates (17.3) and (17.4) are verified.

17.5. Scattering on a wedge. Following [73], we calculate here explicitly the wave operators $W^{\pm} = W^{\pm}(H, H_0; J)$ for the case of the wedge. Thus $\Omega \subset \mathbf{R}^2$ is now defined in the polar coordinates r, θ by the inequalities $0 \leq \theta \leq \theta_0$ where $0 < \theta_0 \leq 2\pi$. In the case $\theta_0 = 2\pi$, the obstacle $\mathcal{O} = \mathbf{R}^2 \setminus \Omega$ reduces to the half-line $x_1 \geq 0$ (to the screen). The operator $H = -\Delta$ is determined by the boundary condition (17.15) where

$$h(x) = h_0|x|^{-1} \quad \text{for} \quad \theta = 0 \quad \text{and} \quad h(x) = h_1|x|^{-1} \quad \text{for} \quad \theta = \theta_0. \tag{17.16}$$

It is assumed that $h_0 \geq 0$, $h_1 \geq 0$; then the Coulomb singularity is not dangerous and the operator H is well-defined, for example, by means of the corresponding quadratic form. The case of the Dirichlet boundary condition is recovered if $h_j = +\infty$ in (17.16).

In the polar coordinates, conditions (17.15), (17.16) mean that

$$\partial u/\partial\theta - h_0 u = 0 \quad \text{for} \quad \theta = 0 \quad \text{and} \quad \partial u/\partial\theta + h_1 u = 0 \quad \text{for} \quad \theta = \theta_0, \tag{17.17}$$

so the variables r and θ can be separated. Indeed, let us consider an auxiliary operator $A = -\partial^2/\partial\theta^2$ with boundary conditions (17.17) in the space $L_2(0,\theta_0)$. The operator A is nonnegative and has the discrete spectrum. Let us denote by ν_m^2, $m = 0, 1, \dots$ and $a_m(\theta)$ its eigenvalues and eigenfunctions. It is easy to see that ν_m satisfy the equation

$$(h_0 + h_1)\cos(\theta_0\nu) - (\nu - h_0h_1\nu^{-1})\sin(\theta_0\nu) = 0. \tag{17.18}$$

The subspace $\mathcal{H}^{(m)}$ of functions

$$f_m(x) = r^{-1/2}g(r)a_m(\theta) \tag{17.19}$$

is invariant with respect to the operator H, and

$$(Hf_m)(x) = r^{-1/2}(H^{(m)}g)(r)a_m(\theta),$$

where the operator

$$H^{(m)} = -d^2/dr^2 + (\nu_m^2 - 1/4)r^{-2}$$

is self-adjoint (if $\nu_m^2 < 1$ the boundary condition $g(0) = 0$ should be added) in the space $L_2(\mathbb{R}_+)$.

The operators $H^{(m)}$ can be diagonalized by the Fourier-Bessel transform. Let $\mathcal{I}_\nu$ be the Bessel function of order ν and

$$(\mathcal{F}_m g)(k) = \int_0^\infty (kr)^{1/2}\mathcal{I}_{\nu_m}(kr)g(r)dr, \quad \nu_m \geq 0.$$

The operators $\mathcal{F}_m$ are unitary in the space $L_2(\mathbb{R}_+)$ and $H^{(m)} = \mathcal{F}_m^* k^2 \mathcal{F}_m$. Therefore, say, for functions $g \in C_0^\infty(\mathbb{R}_+)$,

$$(\exp(-iH^{(m)}t)g)(r) = \int_0^\infty (kr)^{1/2}\mathcal{I}_{\nu_m}(kr)e^{-ik^2t}(\mathcal{F}_m g)(k)dk. \tag{17.20}$$

Using the asymptotics

$$\mathcal{I}_\nu(s) = 2^{1/2}(\pi s)^{-1/2}\cos(s - \pi\nu/2 - \pi/4) + O(s^{-3/2}), \quad s \to \infty,$$

and applying the stationary phase method to integral (17.20), we find that

$$(\exp(-iH^{(m)}t)g)(r) = \mp i(2|t|)^{-1/2}e^{\mp i\pi\nu_m/2}e^{ir^2/(4t)}(\mathcal{F}_m g)(r/(2|t|)) + \varepsilon_m^\pm(r,t), \tag{17.21}$$

where $\varepsilon_m^\pm(\cdot,t)$ tend to zero in $L_2(\mathbb{R}_+)$ as $t \to \pm\infty$.

Let us denote by $\mathcal{F}^\pm$ the operator in

$$\mathcal{H} = L_2(\Omega) = \bigoplus_m \mathcal{H}^{(m)}$$

defined on its every invariant subspace $\mathcal{H}^{(m)}$ by the formula

$$(\mathcal{F}^\pm f_m)(\xi) = e^{\mp i\pi\nu_m/2}k^{-1/2}(\mathcal{F}_m g)(k)a_m(\theta), \quad \xi = k\theta \in \Omega, \tag{17.22}$$

where f_m is function (17.19). Clearly, the operators $\mathcal{F}^{\pm}$ are unitary in $\mathcal{H}$, and it follows from (17.21) that

$$(\exp(-iHt)f)(x) = (2it)^{-1}e^{i|x|^2/(4t)}(\mathcal{F}^{\pm}f)(\pm x/(2t)) + \varepsilon^{\pm}(x,t),$$

where $\varepsilon^{\pm}(\cdot, t)$ tend to zero in $\mathcal{H}$ as $t \to \pm\infty$.

Comparing this expression with the asymptotics (3.3), (3.4) of the free evolution, we see that

$$\lim_{t\to\pm\infty} ||\exp(-iHt)f - J\exp(-iH_0t)f_0^{\pm}|| = 0$$

if (and only if) $(\mathcal{F}^{\pm}f)(\xi) = (\mathcal{F}_0 f_0^{\pm})(\pm\xi)$ for $\xi \in \Omega$. As before, J is the restriction of a function $f \in L_2(\mathbf{R}^2)$ on the wedge Ω, and $\mathcal{F}_0$ is the classical Fourier transform. It follows that the wave operators $W^{\pm}$ exist, are complete and

$$W^+ = (\mathcal{F}^+)^* J\mathcal{F}_0, \quad W^- = (\mathcal{F}^-)^* J\mathcal{J}\mathcal{F}_0,$$

where $\mathcal{J}$ is the reflection operator, $(\mathcal{J}f)(\xi) = f(-\xi)$ (we use the same notation $\mathcal{J}$ as for the reflection operator (3.18) on the sphere). The operator $W^{\pm}$ is isometric on the subspace $\mathcal{H}_0^{\pm}$ of functions $f_0^{\pm}$ such that $(\mathcal{F}_0 f_0^{\pm})(\xi) = 0$ for $\pm\xi \notin \Omega$.

The scattering operator

$$\mathbf{S} = \mathcal{F}_0^* J^* \mathcal{F}^+ (\mathcal{F}^-)^* J\mathcal{J}\mathcal{F}_0 \tag{17.23}$$

and the scattering matrix $S(\lambda)$ are well-defined as bounded operators in the spaces $\mathcal{H}_0$ and $L_2(\mathbf{S}^{d-1})$, respectively. Moreover, the obvious equalities

$$H_0\mathbf{G}_\alpha = \alpha^{-2}\mathbf{G}_\alpha H_0, \quad H\mathbf{G}_\alpha = \alpha^{-2}\mathbf{G}_\alpha H,$$

where the dilation operator $\mathbf{G}_\alpha$ is defined by (3.13), imply that the scattering matrix $S(\lambda) = S$ does not depend on the energy $\lambda > 0$. However, since $\mathcal{H}_0^+ \neq \mathcal{H}_0^-$ (except for the case $\theta_0 = 2\pi$), the scattering operator $\mathbf{S}$ and the scattering matrix S are not unitary. These facts can also be seen from representation (17.23). Therefore it is more natural to consider the modified scattering operator $\mathbf{\Sigma} = \mathbf{S}\mathcal{J}$ and the modified scattering matrix $\Sigma = S\mathcal{J}$ introduced in subsection 3.6. The operator

$$\mathbf{\Sigma} = \mathcal{F}_0^* J^* \mathcal{F}^+ (\mathcal{F}^-)^* J\mathcal{F}_0$$

also is not unitary on $\mathcal{H}_0$, but it is unitary on $\mathcal{H}_0^+$ and it equals zero on the orthogonal complement to it. Consequently, the operator Σ is unitary on the space $L_2(0, \theta_0)$. According to (17.22) $\mathcal{H}^{(m)}$ are invariant subspaces of the operator $\mathcal{F}^+(\mathcal{F}^-)^*$, and it reduces to multiplication by $e^{-i\pi\nu_m}$ on $\mathcal{H}^{(m)}$. Therefore the operator Σ has eigenvalues $e^{-i\pi\nu_m}$ with the corresponding eigenfunctions $a_m(\theta)$. For every m, the function

$$u_m(x) = (\pi k/2)^{1/2} \mathcal{J}_{\nu_m}(kr) a_m(\theta)$$

satisfies in Ω the Helmholtz equation $-\Delta u = \lambda u$, the boundary conditions (17.17) and has the asymptotics of the standing wave (3.20) as $|x| \to \infty$.

Let us consider the case of the Dirichlet and Neumann boundary conditions. It follows from (17.18) that $\nu_m = \pi(m+1)\theta_0^{-1}$ in the first case and $\nu_m = \pi m\theta_0^{-1}$ in the second. If $\theta_0\pi^{-1}$ is rational, then in both cases the operator Σ has only a finite number of eigenvalues (of infinite multiplicity each). In particular, Σ is the identity operator if $\theta_0\pi^{-1} = (2p)^{-1}$, $p = 1, 2, \ldots$. If $\theta_0\pi^{-1}$ is irrational, then the point spectrum of the operator Σ is dense on the unit circle.

In the case of the screen when $\theta_0 = 2\pi$, the spaces $\mathcal{H}_0$ and $\mathcal{H}$ coincide, $J = I$ and the operators $W^{\pm}$ are unitary. Both for the Dirichlet and Neumann boundary conditions, the spectrum of Σ consists of the four points ± 1 and $\pm i$. Moreover, for the screen, one can calculate the spectrum of the scattering matrix S (see the original paper [73]). For the Dirichlet (Neumann) boundary condition, it consists of the absolutely continuous part covering the lower (upper) half-circle and of the eigenvalue 1 of infinite multiplicity. Recall that the Dirichlet (Neumann) problem corresponds to a positive (negative) perturbation of the operator H_0. Thus this result qualitatively agrees with that of Theorem 8.3 on the spectrum of S for the Schrödinger operator with a positive (negative) potential. Finally, we note that for the general boundary condition (17.17) the essential spectrum of S is the same as in the Neumann case.

Bibliography

[1] S. Agmon, Spectral properties of Schrödinger operators and scattering theory, Ann. Scuola Norm. Sup. Pisa **2** no. 4 (1975), 151-218.

[2] S. Agmon, Some new results in spectral and scattering theory of differential operators in $\mathbb{R}^n$, Seminaire Goulaouic Schwartz, Ecole Polytechnique, 1978.

[3] S. Agmon, *Lectures on Exponential Decay of Solutions of Second-Order Elliptic Equations*, Math. Notes, Princeton Univ. Press, 1982.

[4] S. Agmon and H. Hörmander, Asymptotic properties of solutions of differential equations with simple characteristics, J. Anal. Math. **30** (1976), 1-38.

[5] W. O. Amrein, A. Boutet de Monvel and V. Georgescu, *C_0-Groups, Commutator Methods and Spectral Theory for N-Body Hamiltonians*, Progress in Math. Physics. Press, **135**, Birkhäuser, 1996.

[6] W. O. Amrein, J. M. Jauch and K. B. Sinha, *Scattering theory in quantum mechanics*, Benjamin, New York, 1977.

[7] W. O. Amrein, D. B. Pearson and K. B. Sinha, Bounds on the total cross section for N-body systems, Nuovo Cimento **52 A** (1979), 115-131.

[8] H. Baumgärtel and M. Wollenberg, *Mathematical scattering theory*, Akademie-Verlag, Berlin, 1983.

[9] Yu. M. Berezanskii, *Expansion in eigenfunctions of selfadjoint operators*, Amer. Math. Soc., Providence, R.I., 1968.

[10] M. Sh. Birman, Perturbation of the continuous spectrum of a singular elliptic operator under the change of the boundary and boundary conditions, Vestnik Leningrad Univ. Math. **1** (1962), 22-55 (Russian).

[11] M. Sh. Birman, Existence conditions for wave operators, Izv. Akad. Nauk SSSR, Ser. Mat. **27** no. 4 (1963), 883-906 (Russian).

[12] M. Sh. Birman, A local criterion for the existence of wave operators, Soviet Math. Dokl. **5** no. 2 (1965), 1505-1509.

[13] M. Sh. Birman, A local test for the existence of wave operators, Math. USSR-Izv. **2** no. 2 (1968), 879-906.

[14] M. Sh. Birman, Scattering problems for differential operators with constant coefficients, Funct. Anal. Appl. **3** no. 3 (1969), 167-180.

[15] M. Sh. Birman and S. B. Entina, Stationary approach in abstract scattering theory, Math. USSR-Izv. **1** no. 1 (1967), 391-420.

[16] M. Sh. Birman and M. G. Kreĭn, On the theory of wave operators and scattering operators, Soviet Math. Dokl. **3** (1962), 740-744.

[17] M. Sh. Birman and M. G. Kreĭn, Some topics of the theory of wave and scattering operators, Outlines Joint Soviet-Amer. Sympos. Part. Diff. Equations, Novosibirsk, 1963, 39-45.

[18] M. Sh. Birman and A. B. Pushnitski, Spectral shift function, amazing and multifaceted, Int. Eq. Op. Theory **30** (1998), 191-199.

[19] M. Sh. Birman and M. Z. Solomyak, Asymptotic behavior of the spectrum of pseudodifferential operators with anisotropically homogeneous symbols, I, II, Vestnik Leningrad Univ. Math. **10** (1980), 237-247 and **12** (1982), 155-161.

[20] M. Sh. Birman and D. R. Yafaev, Asymptotics of the spectrum of the scattering matrix, J. Soviet Math. **25** no. 2 (1984), 793-814.

[21] M. Sh. Birman and D. R. Yafaev, A general scheme in the stationary theory of scattering, Problemy Math. Phys. **12** (1987), 89-117. English trasl., Amer. Math. Soc. Transl. (Ser.2) **157** (1993), 87-112.

[22] M. Sh. Birman and D. R. Yafaev, On the trace-class method in potential scattering theory, J. Soviet Math. **56** no. 2 (1991), 2285-2299.

[23] M. Sh. Birman and D. R. Yafaev, The spectral shift function. The papers of M. G. Kreĭn and their further development, St. Petesburg Math. J. **4** no. 5 (1993), 833-870.

[24] M. Sh. Birman and D. R. Yafaev, Spectral properties of the scattering matrix, St. Petesburg Math. J. **4** no. 6 (1993), 1055-1079.

[25] M. Sh. Birman and D. R. Yafaev, The scattering matrix for a perturbation of a periodic Schrödinger operator by decreasing potential, St. Petesburg Math. J. **6** no. 3 (1995), 453-474.

[26] V. S. Buslaev, Trace formulas and certain asymptotic estimates of the resolvent kernel for the Schrödinger operator in three-dimensional space, in Topics in Math. Phys. **1**, Plenum Press, 1967.

[27] V. S. Buslaev, The trace formulas and the singularities of the scattering matrix for the system of three one-dimensional particles, Teoret. Matem. Fiz. **16**, No.2 (1973), 247-259 (Russian).

[28] V. S. Buslaev and N. A. Kaliteevsky, Principal singularities of the scattering matrix for a system of one-dimensional particles, Teoret. Matem. Fiz. **70**, No.2 (1987), 266-277 (Russian).

[29] V. S. Buslaev and V. B. Matveev, Wave operators for the Schrödinger equation with a slowly decreasing potential, Theor. Math. Phys. **2** (1970), 266-274.

[30] V. S. Buslaev and S. P. Merkur'ev, On the connection between the third virial coefficient and the scattering matrix, Teoret. Matem. Fiz. **5**, No.3 (1970), 372-387 (Russian).

[31] V. S. Buslaev, S. P. Merkur'ev and S. P. Salikov, On the diffraction character of scattering in a quantum system of three one-dimensional particles, in Topics in Math. Phys. **9**, Plenum Press, 1979; English transl., Sel. Math. Sov. **6** (1987), 45-58.

[32] P. Constantin, Scattering for Schrödinger operators in a class of domains with noncompact boundaries, J. Funct. Anal. **44** (1981), 87-119.

[33] J. M. Cook, Convergence to the Møller wave matrix, J. Math. Phys. **36** (1957), 82-87.

[34] H. Cycon, R. Froese, W. Kirsch and B. Simon, *Schrödinger operators*, Texts and Monographs in Physics, Springer-Verlag, Berlin, Heidelberg, New York, 1987.

[35] V. G. Deich, E. L. Korotyaev and D. R. Yafaev, Theory of potential scattering taking into account spatial anisotropy, J. Soviet Math. **34** (1986), 2040-2050.

[36] P. Deift, Classical scattering theory with a trace class condition, Princeton Univ Press, Princeton, New Jersey, 1979.

[37] P. Deift and B. Simon, A time-dependent approach to the completeness of multiparticle quantum systems, Comm. Pure Appl. Math. **30** (1977), 573-583.

[38] J. Dereziński, The Mourre estimate for dispersive N-body Schrödinger operators, Trans. Amer. Math. Soc. **317** (1990), 773-798.

[39] J. Dereziński, Asymptotic completeness of long-range quantum systems, Ann. Math. **138** (1993), 427-473.

[40] J. Dereziński and C. Gérard, A remark on the asymptotic clustering of N-body systems, Lecture Notes in Phys. **403**, Springer-Verlag (1992), 73-78.

[41] J. Dereziński and C. Gérard, *Scattering theory of classical and quantum N particle systems*, Springer-Verlag, 1997.

[42] J. Dollard, Asymptotic convergence and Coulomb interaction, J. Math. Phys. **5** (1964), 723-738.

[43] D. M. Eidus and A. A. Vinnik, The radiation conditions for domains with infinite boundaries, Dokl. Akad. Nauk SSSR **214** no.2 (1974), 19-21. (Russian)

[44] V. Enss, Completeness of three-body quantum scattering, in: *Dynamics and processes*, P. Blanchard and L. Streit, eds., Springer Lecture Notes in Math. **1031** (1983), 62-88.

[45] V. Enss, Quantum scattering theory for two- and three-body systems with potentials of short and long range, in: *Schrödinger operators*, S. Graffi, ed., Springer Lecture Notes in Math. **1159** (1985), 39-176.

[46] V. Enss, Long-range scattering of two- and three-body quantum systems, in: *Journées EDP,* Saint Jean de Monts (1989), 1-31.

[47] V. Enss, B. Simon, Finite total cross sections in nonrelativistic quantum mechanics, Comm. Math. Phys. **76** (1980), 177-209.

[48] L. D. Faddeev, *Mathematical Aspects of the Three Body Problem in Quantum Scattering Theory*, Israel Program of Sci. Transl., 1965.

[49] L. D. Faddeev, On the Friedrichs model in the theory of perturbations of the continuous spectrum, Amer. Math. Soc. Transl. (Ser.2) **62** (1967), 177-203.

[50] L. D. Faddeev and S. P. Merkur'ev, Quantum scattering theory for several particles systems, MPAM No. 11, Kluver Academic Press Publishers, 1993.

[51] L. D. Faddeev and O. A. Yakubovski, *Lectures on quantum mechanics for mathematics students*, Izdat. Leningrad. Univ., Leningrad, 1980. (Russian)

[52] K. Friedrichs, On the perturbation of the continuous spectra, Comm. Pure Appl. Math. **1** (1948), 361-406.

[53] K. Friedrichs, *Perturbation of spectra in Hilbert space*, Amer. Math. Soc., Providence, R.I., 1965.

[54] R. Froese, I. Herbst, A new proof of the Mourre estimate, Duke Math. J. **49** (1982), 1075-1085.

[55] Y. Gâtel and D. R. Yafaev, On solutions of the Schrödinger equation with radiation conditions at infinity: the long-range case, Ann. Institut Fourier **49**, no. 5 (1999), 1581-1602.

[56] C. Gérard, The Mourre estimate for regular dispersive systems, Ann. I. H. P. **54** (1991), 59-88.

[57] C. Gérard, Asymptotic completeness of 3-particle long-range systems, Invent. Math. **114** (1993), 333-397.

[58] C. Gérard, F. Nier, Scattering theory for the perturbations of periodic Schrödinger operators, J. Math. Kyoto Univ. **38** (1998), 595-634.

[59] J. Ginibre and M. Moulin, Hilbert space approach to the quantum mechanical three body problem, Ann. Inst. H.Poincaré, A **21** (1974), 97-145.

[60] G. M. Graf, Asymptotic completeness for N-body short-range quantum systems: A new proof, Comm. Math. Phys. **132** (1990), 73-101.

[61] G. M. Graf, D. Schenker, 2-Magnon scattering in the Heizenberg model, Ann. Inst. H.Poincaré, A **67** (1997), 91-107.

[62] K. Hepp, On the quantum mechanical N-body problem, Helv. Phys. Acta **42** (1969), 425-458.

[63] I. Herbst, On the connectedness structure of the Coulomb S-matrix, Comm. Math. Phys. **35** (1974), 181-191.

[64] I. Herbst, E. Skibsted, Free channel Fourier transform in the long-range N-body problem, J. Analyse Math. **65** (1995), 297-332.

[65] L. Hörmander, Lower bounds at infinity for solutions of differential equations with constant coefficients, Israel J. Math. **16** no.1 (1973), 103-116.

[66] L. Hörmander, The existence of wave operators in scattering theory, Math. Z. **146** (1976), 69-91.

[67] L. Hörmander, *The analysis of linear partial differential operators* III, IV, Springer-Verlag, 1985.

[68] W. Hunziker, I. M. Sigal, Time-dependent scattering theory of N-body quantum systems, Rev. Math. Phys. to appear.

[69] T. Ikebe, Eigenfunction expansions associated with the Schrödinger operators and their application to scattering theory, Arch. Rational Mech. Anal. **5** (1960), 1-34; Erratum, Remarks on the orthogonality of eigenfunctions for the Schrödinger operator on $\mathbf{R}^n$, J. Fac. Sci. Univ. Tokyo Sect. I **17** (1970), 355-361.

[70] T. Ikebe, Spectral Representation for Schrödinger Operators with Long-Range Potentials, J. Funct. Anal. **20** (1975), 158-177.

[71] T. Ikebe, H. Isozaki, A stationary approach to the existence and completeness of long-range wave operators, Int. Eq. Op. Theory **5** (1982), 18-49.

[72] T. Ikebe, Y. Saito, Limiting absorption method and absolute continuity for the Schrödinger operator, J. Math. Kyoto Univ. **12** (1972), 513-542.

[73] E. M. Il'in, Scattering matrix for the diffraction problem on a wedge, Operator theory and function theory, LGU, **1** (1983), 87-100. (Russian)

[74] E. M. Il'in, The limiting absorption principle and scattering by noncompact obstacles, Soviet Math. (Izv. VUZ) **28** (1984).

[75] E. M. Il'in, Scattering by unbounded obstacles for elliptic operators of second order, Proc. of the Steklov Inst. of Math. **179** (1989), 85-107.

[76] R. J. Iorio and M. O'Carrol, Asymptotic completeness for multi-particle Schrödinger Hamiltonians with weak potentials, Comm. Math. Phys. **27** (1972), 137-145.

[77] H. Isozaki, Eikonal equations and spectral representations for long range Schrödinger Hamiltonians, J. Math. Kyoto Univ. **20** (1980), 243-261.

[78] H. Isozaki, Structures of S-matrices for three body Schrödinger operators, Comm. Math. Phys. **146** (1992), 241-258.

[79] H. Isozaki, Asymptotic properties of generalized eigenfunctions for three body Schrödinger operators, Comm. Math. Phys. **153** (1993), 1-21.

[80] H. Isozaki, On N-body Schrödinger operators, Proc. Indian Acad. Sci. **104**, No. 4 (1994), 667-703.

[81] H. Isozaki, H. Kitada, Micro-local resolvent estimates for 2-body Schrödinger operators, J. Funct. Anal. **57** (1984), 270-300.

[82] H. Isozaki, H. Kitada, Modified wave operators with time-independent modifies, J. Fac. Sci, Univ. Tokyo, **32** (1985), 77-104.

[83] H. Isozaki, H. Kitada, Scattering matrices for two-body Schrödinger operators, Sci. Papers College Arts and Sci., Univ. Tokyo, **35** (1985), 81-107.

[84] H. Isozaki, H. Kitada, A remark on the micro-local resolvent estimates for two-body Schrödinger operators, Publ. RIMS, Kyoto Univ., **21** (1986), 889-910.

[85] W. Jäger, Zur Theorie der Schwingungsgleichung mit variablen Koeffizienten in Aussengebieten, Math. Z. **102** (1967), 62-89.

[86] W. Jäger, Das asymptotische Verhalten von Lösungen eines Types von Differentialgleichungen, Math. Z. **112** (1969), 26-37.

[87] W. Jäger, Ein gewöhnlicher Differentialoperator zweiter Ordnung für Funktionen mit Werten in einem Hilbertraum, Math. Z. **113** (1970), 68-98.

[88] A. Jensen, Propagation estimates for Schrödinger-type operators, Trans. Amer. Math. Soc. **291** (1985), 129-144.

[89] A. Jensen, T. Kato, Spectral properties of Schrödinger operators and time decay of the wave functions, Duke Math. J. **46** (1979), 583-611.

[90] A. Jensen, E. Mourre, P. Perry, Multiple commutator estimates and resolvent smoothness in quantum scattering theory, Ann. Inst. Henri Poincaré, ph. th. **41** (1984), 207-225.

[91] A. Jensen, P. Perry, Commutator Methods and Besov Space estimates for Schrödinger operators, J. Operator Theory, **14** (1985), 181-188.

[92] T. Kato, On finite-dimensional perturbations of self-adjoint operators, J. Math. Soc. Japan. **9** (1957), 239-249.

[93] T. Kato, Perturbations of the continuous spectra by trace class operators, Proc. Japan. Acad. Ser.A. Math. Sci. **33** (1957), 260-264.

[94] T. Kato, Growth properties of solutions of the reduced wave equation with variable coefficients, Comm. Pure Appl. Math. **12** (1959), 403-425.

[95] T. Kato, Wave operators and unitary equivalence, Pacific J. Math. **15** (1965), 171-180.

[96] T. Kato, Wave operators and similarity for some non-self-adjoint operators, Math. Ann. **162** (1966), 258-279.

[97] T. Kato, *Perturbation theory for linear operators*, Springer-Verlag, 1966.

[98] T. Kato, Scattering theory with two Hilbert spaces, J. Funct. Anal. **1** (1967), 342-369.

[99] T. Kato, Smooth operators and commutators, Studia Math. **31**(1968), 535-546.

[100] T. Kato, Some results on potential scattering, In: Proc. Intern. Conf. on Funct. Anal. and Related Topics, 1969, Univ. of Tokyo Press, Tokyo, 206-215.

[101] T. Kato, Monotonicity theorems in scattering theory, Hadronic J. **1**(1978), 134-154.

[102] E. L. Korotyaev and D. R. Yafaev, Traces on surfaces for function classes with dominant mixed derivatives, J. Soviet Math. **8**, no. 1 (1978), 73-86.

[103] M. G. Kreĭn, On the trace formula in perturbation theory, Mat. Sb. **33** (1953), 597-626. (Russian)

[104] M. G. Kreĭn, On perturbation determinants and the trace formula for unitary and selfadjoint operators, Soviet Math. Dokl. **3** (1962).

[105] M. G. Kreĭn, Some new studies in the theory of perturbations of selfadjoint operaors, First Math. Summer School (Kanev, 1963) "Naukova Dumka", Kiev, 1964, 103-187; English transl. in: M. G. Kreĭn, Topics in differential and integral equations and operator theory, Birkhäser, Basel, 1983, 107-172.

[106] S. T. Kuroda, On the existence and the unitarity of the scattering operator, Nuovo Cimento **12** (1959), 431-454.

[107] S. T. Kuroda, Perturbation of continuous spectra by unbounded operators. I, J. Math. Soc. Japan **11** (1959), 247-262; II, J. Math. Soc. Japan **12** (1960), 243-257.

[108] S. T. Kuroda, On the Hölder continuity of an integral involving Bessel functions, Quart. J. Math. Oxford **21**, no. 2 (1970), 71-81.

[109] S. T. Kuroda, Some remarks on scattering theory for Schrödinger operators, J. Fac. Sci., Univ. Tokyo, ser. I **17** (1970), 315-329.

[110] S. T. Kuroda, Scattering theory for differential operators, J. Math. Soc. Japan **25** no. 1,2 (1973), 75-104, 222-234.

[111] L. D. Landau and E. M. Lifshitz, *Classical mechanics*, Pergamon Press, 1960.

[112] L. D. Landau and E. M. Lifshitz, *Quantum mechanics*, Pergamon Press, 1965.

[113] R. Lavine, Commutators and scattering theory I: Repulsive interactions, Comm. Math. Phys. **20**(1971), 301-323; II:A class of one-body problems, Indiana Univ. Math. J. **21**(1972), 643-656.

[114] R. Lavine, Completeness of the wave operators in the repulsive N-body problem, J. Math. Phys. **14** (1973), 376-379.

[115] N. Lerner, D. Yafaev, Trace theorems for pseudo-differential operators, J. Analyse Math. **74** (1998), 113-164.

[116] I. M. Lifshits, On a problem in perturbation theory, Uspehi Mat. Nauk **7** (1952), no. 1, 171-180 (Russian).

[117] R. Melrose, M. Zworski, Scattering metrics and geodesic flow at infinity, Invent. Math. **124** (1996), 389-436.

[118] S. P. Mercur'ev, Coordinate asymptotics of the wave function for three-particle systems, Teoret. Matem. Fiz. **8** (1971), 235-250 (Russian).

[119] C. Møller, General properties of the characteristic matrix in the theory of elementary particles.I, Danske Vid. Selsk. Mat.-Fyz. Medd. **23** (1945), no. 1, 1-48.

[120] E. Mourre, Absence of singular spectrum for certain self-adjoint operators, Comm. Math. Phys. **78** (1981), 391-400.

[121] E. Mourre, Opérateurs conjugués et propriétés de propagation, Comm. Math. Phys. **91** (1983), 279-300.

[122] R. Newton, Fredholm methods in the three-body problem. I, J. Math. Phys. **12** (1971), 1552-1567.

[123] R. Newton, The three particle S-matrix, J. Math. Phys. **15** (1974), 338-343.

[124] R. Newton, Noncentral potentials: The generalized Levinson theorem and the structure of the spectrum, J. Math. Phys. **18** (1977), 1348-1357.

[125] R. Newton, *Inverse Schrödinger scattering in three dimensions*, Springer-Verlag, 1989.

[126] J. Nuttal, Asymptotic form of the three-particle scattering wave function for free incident particles, J. Math. Phys. **12** (1971), 1896-1899.

[127] D. B. Pearson, An example in potential scattering illustrating the breakdown of asymptotic completeness, Comm. Math. Phys. **40** (1975), 125-146.

[128] D. B. Pearson, A generalization of the Birman trace theorem, J. Funct. Anal. **28** (1978), 182-186.

[129] D. B. Pearson, Quantum scattering and spectral theory, Academic Press, 1988.

[130] P. Perry, I. M. Sigal and B. Simon, Spectral analysis of N-body Schrödinger operators, Ann. Math. **144** (1981), 519-567.

[131] A. Ya. Povzner, On the expansion of arbitrary functions in the eigenfunctions of the operator $-\Delta u + qu$, Mat. Sb. **32** (1953), no.1, 109-156. (Russian)

[132] A. Ya. Povzner, On expansion in eigenfunctions of the Schrödinger equation, Dokl. Akad. Nauk SSSR **104** (1955), no.3, 360-363. (Russian)

[133] C. R. Putnam, *Commutator properties of Hilbert space operators and related topics*, Springer-Verlag, Berlin, Heidelberg, New York, 1967.

[134] M. Reed and B. Simon, *Methods of Modern Mathematical Physics*, Academic Press, II, 1975; III, 1979; IV, 1978.

[135] M. Rosenblum, Perturbation of the continuous spectrum and unitary equivalence, Pacific J. Math. **7** (1957), 997-1010.

[136] Y. Saito, *Spectral Representation for Schrödinger Operators with Long-Range Potentials*, Springer Lecture Notes in Math. **727**, 1979.

[137] Y. Saito, On the S-matrix for Schrödinger operators with long-range potentials, J. reine Angew. Math. **314** (1980), 99-116.

[138] M. A. Shubin, *Pseudodifferential Operators and Spectral Theory*, Springer-Verlag, 1987.

[139] I. M. Sigal, Mathematical Foundations of Quantum Scattering Theory for Multiparticle Systems, Memoirs of the Amer. Math. Soc. **N 209**, 1978.

[140] I. M. Sigal and A. Soffer, The N-particle scattering problem: Asymptotic completeness for short-range systems, Ann. Math. **126** (1987), 35-108.

[141] I. M. Sigal and A. Soffer, Long-range many-body scattering. Asymptotic clustering for Coulomb-type potentials, Invent. Math. **99** (1990), 115-143.

[142] I. M. Sigal and A. Soffer, Asymptotic completeness for four-body Coulomb systems, Duke Math. J. **71** (1993), 243-298.

[143] I. M. Sigal and A. Soffer, Asymptotic completeness of N-particle long range systems, Journal AMS **7** (1994), 307-333.

[144] E. Skibsted, Smoothness of N-body scattering amplitudes, Rev. Math. Phys. **4**, no. 4 (1992), 619-658.

[145] M. M. Skriganov, Uniform coordinate and spectral asymptotics for solutions of the scattering problem for the Schrödinger equation, J. Soviet Math. **8**, no. 1, 120- 141.

[146] A. V. Sobolev and D. R. Yafaev, On the quasiclassical limit of the total scattering cross section in nonrelativistic quantum mechanics, Ann. Inst. H.Poincaré, ph.th., **44** (1986), 195-210.

[147] A. V. Sobolev and D. R. Yafaev, Spectral properties of the abstract scattering matrix, Proc. Steklov Inst. Math. **188**, no. 3 (1991), 159-189.

[148] L. E. Thomas, Time dependent approach to scattering from impurities in a crystal, Comm. Math. Phys. **33** (1973), 335-343.

[149] L. E. Thomas, Asymptotic completeness in two- and three-particle quantum mechanical scattering, Ann. Phys. **90** (1975), 127-165.

[150] A. Vasy, Scattering matrices in many-body scattering, Comm. Math. Phys. **200** (1999), 105-124.

[151] A. A. Vinnik, Radiation conditions for domains with infinite boundaries, Soviet Math. (Izv. VUZ) **21** (1977).

[152] V. Vogelsang, Das Ausstrahlungsproblem für elliptische Differentialgleichungen in Gebieten mit unbeschränkten Rand, Math. Z. **144** (1975), 101-125.

[153] X. P. Wang, On the three body long range scattering problems, Reports Math. Phys. **25**(1992), 267-276.

[154] D. R. Yafaev, The virtual level of the Schrödinger equation, Zap. Naučhn. Sem. LOMI **51** (1975), 203-216. English transl., J. Sov. Math. **11** (1979), 501-510.

[155] D. R. Yafaev, On the multichannel scattering theory in two spaces, Theor. Math. Phys. **37** (1978), 867-874.

[156] D. R. Yafaev, On the singular spectrum in a system of three particles, Math. USSR-Sb. **35** (1979), 283-300.

[157] D. R. Yafaev, On the break-down of completeness of wave operators in potential scattering, Comm. Math. Phys. **65** (1979), 167-179.

[158] D. R. Yafaev, Wave operators for the Schrödinger equation, Theoret. and Math. Phys. **45**, no. 2 (1981), 992-998.

[159] D. R. Yafaev, Scattering theory for time-dependent zero-range potentials, Ann. Inst. H.Poincaré, ph.th., **40** (1984), 343-359.

[160] D. R. Yafaev, Remarks on the spectral theory for the multiparticle type Schrödinger operator, J. Soviet Math. **31** (1985), 3445-3459.

[161] D. R. Yafaev, On a resonant scattering on a negative potential, J. Soviet Math. **32**, no. 5 (1986), 549-556.

[162] D. R. Yafaev, On the asymptotics of the scattering phases for the Schrödinger equation, Ann. Inst. H.Poincaré, ph.th., **53** (1990), 283-299.

[163] D. R. Yafaev, On solutions of the Schrödinger equation with radiation conditions at infinity, Advances Sov. Math. **7** (1991), 179-204.

[164] D. R. Yafaev, On the scattering matrix for perturbations of constant sign, Ann. Inst. H.Poincaré, ph.th., **57** (1992), 361-384.

[165] D. R. Yafaev, *Mathematical Scattering Theory*, Amer. Math. Soc., 1992.

[166] D. R. Yafaev, Radiation conditions and scattering theory for N-particle Hamiltonians, Comm. Math. Phys. **154** (1993), 523-554.

[167] D. R. Yafaev, Eigenfunctions of the continuous spectrum for the N-particle Schrödinger operator, in: *Spectral and scattering theory*, M. Ikawa, ed., Marcel Dekker, Inc. (1994), 259-286.

[168] D. R. Yafaev, Resolvent estimates and scattering matrix for N-particle Hamiltonians, Int. Eq. Op. Theory **21** (1995), 93-126.

[169] D. R. Yafaev, New channels of scattering for three-body quantum systems with long-range potentials, Duke Math. J. **82** (1996), 553-584.

[170] D. R. Yafaev, New channels in the two-body long-range scattering, St. Petersburg Math. J. **8** (1997), 165-182.

[171] D. R. Yafaev, On the classical and quantum Coulomb scattering, J. Phys. A: Math. Gen. **30** (1997), 6981-6992.

[172] D. R. Yafaev, The scattering amplitude for the Schrödinger equation with a long-range potential, Comm. Math. Phys. **191** (1998), 183-218.

[173] D. R. Yafaev, Scattering theory: some old and new problems, Documenta Math., Proc. of the ICM, Berlin 1998, v. 3, 87-96.

[174] D. R. Yafaev, A class of pseudo-differential operators with oscillating symbols, St. Petersburg Math. J. **11** no. 2 (2000).

[175] O. A. Yakubovsky, On the integral equations in the theory of N-particle scattering, Soviet J. Nuclear Phys. **5** (1967), 937-942.

Index

Druck: Strauss Offsetdruck, Mörlenbach
Verarbeitung: Schäffer, Grünstadt

Vol. 1587: N. Xi, Representations of Affine Hecke Algebras. VIII, 137 pages. 1994.

Vol. 1588: C. Scheiderer, Real and Étale Cohomology. XXIV, 273 pages. 1994.

Vol. 1589: J. Bellissard, M. Degli Esposti, G. Forni, S. Graffi, S. Isola, J. N. Mather, Transition to Chaos in Classical and Quantum Mechanics. Montecatini Terme, 1991. Editor: 2S. Graffi. VII, 192 pages. 1994.

Vol. 1590: P. M. Soardi, Potential Theory on Infinite Networks. VIII, 187 pages. 1994.

Vol. 1591: M. Abate, G. Patrizio, Finsler Metrics – A Global Approach. IX, 180 pages. 1994.

Vol. 1592: K. W. Breitung, Asymptotic Approximations for Probability Integrals. IX, 146 pages. 1994.

Vol. 1593: J. Jorgenson & S. Lang, D. Goldfeld, Explicit Formulas for Regularized Products and Series. VIII, 154 pages. 1994.

Vol. 1594: M. Green, J. Murre, C. Voisin, Algebraic Cycles and Hodge Theory. Torino, 1993. Editors: A. Albano, F. Bardelli. VII, 275 pages. 1994.

Vol. 1595: R.D.M. Accola, Topics in the Theory of Riemann Surfaces. IX, 105 pages. 1994.

Vol. 1596: L. Heindorf, L. B. Shapiro, Nearly Projective Boolean Algebras. X, 202 pages. 1994.

Vol. 1597: B. Herzog, Kodaira-Spencer Maps in Local Algebra. XVII, 176 pages. 1994.

Vol. 1598: J. Berndt, F. Tricerri, L. Vanhecke, Generalized Heisenberg Groups and Damek-Ricci Harmonic Spaces. VIII, 125 pages. 1995.

Vol. 1599: K. Johannson, Topology and Combinatorics of 3-Manifolds. XVIII, 446 pages. 1995.

Vol. 1600: W. Narkiewicz, Polynomial Mappings. VII, 130 pages. 1995.

Vol. 1601: A. Pott, Finite Geometry and Character Theory. VII, 181 pages. 1995.

Vol. 1602: J. Winkelmann, The Classification of Three-dimensional Homogeneous Complex Manifolds. XI, 230 pages. 1995.

Vol. 1603: V. Ene, Real Functions – Current Topics. XIII, 310 pages. 1995.

Vol. 1604: A. Huber, Mixed Motives and their Realization in Derived Categories. XV, 207 pages. 1995.

Vol. 1605: L. B. Wahlbin, Superconvergence in Galerkin Finite Element Methods. XI, 166 pages. 1995.

Vol. 1606: P.-D. Liu, M. Qian, Smooth Ergodic Theory of Random Dynamical Systems. XI, 221 pages. 1995.

Vol. 1607: G. Schwarz, Hodge Decomposition – A Method for Solving Boundary Value Problems. VII, 155 pages. 1995.

Vol. 1608: P. Biane, R. Durrett, Lectures on Probability Theory. Editor: P. Bernard. VII, 210 pages. 1995.

Vol. 1609: L. Arnold, C. Jones, K. Mischaikow, G. Raugel, Dynamical Systems. Montecatini Terme, 1994. Editor: R. Johnson. VIII, 329 pages. 1995.

Vol. 1610: A. S. Üstünel, An Introduction to Analysis on Wiener Space. X, 95 pages. 1995.

Vol. 1611: N. Knarr, Translation Planes. VI, 112 pages. 1995.

Vol. 1612: W. Kühnel, Tight Polyhedral Submanifolds and Tight Triangulations. VII, 122 pages. 1995.

Vol. 1613: J. Azéma, M. Emery, P. A. Meyer, M. Yor (Eds.), Séminaire de Probabilités XXIX. VI, 326 pages. 1995.

Vol. 1614: A. Koshelev, Regularity Problem for Quasilinear Elliptic and Parabolic Systems. XXI, 255 pages. 1995.

Vol. 1615: D. B. Massey, Le Cycles and Hypersurface Singularities. XI, 131 pages. 1995.

Vol. 1616: I. Moerdijk, Classifying Spaces and Classifying Topoi. VII, 94 pages. 1995.

Vol. 1617: V. Yurinsky, Sums and Gaussian Vectors. XI, 305 pages. 1995.

Vol. 1618: G. Pisier, Similarity Problems and Completely Bounded Maps. VII, 156 pages. 1996.

Vol. 1619: E. Landvogt, A Compactification of the Bruhat-Tits Building. VII, 152 pages. 1996.

Vol. 1620: R. Donagi, B. Dubrovin, E. Frenkel, E. Previato, Integrable Systems and Quantum Groups. Montecatini Terme, 1993. Editors:M. Francaviglia, S. Greco. VIII, 488 pages. 1996.

Vol. 1621: H. Bass, M. V. Otero-Espinar, D. N. Rockmore, C. P. L. Tresser, Cyclic Renormalization and Auto-morphism Groups of Rooted Trees. XXI, 136 pages. 1996.

Vol. 1622: E. D. Farjoun, Cellular Spaces, Null Spaces and Homotopy Localization. XIV, 199 pages. 1996.

Vol. 1623: H.P. Yap, Total Colourings of Graphs. VIII, 131 pages. 1996.

Vol. 1624: V. Brınzanescu, Holomorphic Vector Bundles over Compact Complex Surfaces. X, 170 pages. 1996.

Vol.1625: S. Lang, Topics in Cohomology of Groups. VII, 226 pages. 1996.

Vol. 1626: J. Azéma, M. Emery, M. Yor (Eds.), Séminaire de Probabilités XXX. VIII, 382 pages. 1996.

Vol. 1627: C. Graham, Th. G. Kurtz, S. Méléard, Ph. E. Protter, M. Pulvirenti, D. Talay, Probabilistic Models for Nonlinear Partial Differential Equations. Montecatini Terme, 1995. Editors: D. Talay, L. Tubaro. X, 301 pages. 1996.

Vol. 1628: P.-H. Zieschang, An Algebraic Approach to Association Schemes. XII, 189 pages. 1996.

Vol. 1629: J. D. Moore, Lectures on Seiberg-Witten Invariants. VII, 105 pages. 1996.

Vol. 1630: D. Neuenschwander, Probabilities on the Heisenberg Group: Limit Theorems and Brownian Motion. VIII, 139 pages. 1996.

Vol. 1631: K. Nishioka, Mahler Functions and Transcendence. VIII, 185 pages. 1996.

Vol. 1632: A. Kushkuley, Z. Balanov, Geometric Methods in Degree Theory for Equivariant Maps. VII, 136 pages. 1996.

Vol.1633: H. Aikawa, M. Essén, Potential Theory – Selected Topics. IX, 200 pages. 1996.

Vol. 1634: J. Xu, Flat Covers of Modules. IX, 161 pages. 1996.

Vol. 1635: E. Hebey, Sobolev Spaces on Riemannian Manifolds. X, 116 pages. 1996.

Vol. 1636: M. A. Marshall, Spaces of Orderings and Abstract Real Spectra. VI, 190 pages. 1996.

Vol. 1637: B. Hunt, The Geometry of some special Arithmetic Quotients. XIII, 332 pages. 1996.

Vol. 1638: P. Vanhaecke, Integrable Systems in the realm of Algebraic Geometry. VIII, 218 pages. 1996.

Vol. 1639: K. Dekimpe, Almost-Bieberbach Groups: Affine and Polynomial Structures. X, 259 pages. 1996.

Vol. 1640: G. Boillat, C. M. Dafermos, P. D. Lax, T. P. Liu, Recent Mathematical Methods in Nonlinear Wave Propagation. Montecatini Terme, 1994. Editor: T. Ruggeri. VII, 142 pages. 1996.

Vol. 1641: P. Abramenko, Twin Buildings and Applications to S-Arithmetic Groups. IX, 123 pages. 1996.

Vol. 1642: M. Puschnigg, Asymptotic Cyclic Cohomology. XXII, 138 pages. 1996.

Vol. 1643: J. Richter-Gebert, Realization Spaces of Polytopes. XI, 187 pages. 1996.

Vol. 1644: A. Adler, S. Ramanan, Moduli of Abelian Varieties. VI, 196 pages. 1996.

Vol. 1645: H. W. Broer, G. B. Huitema, M. B. Sevryuk, Quasi-Periodic Motions in Families of Dynamical Systems. XI, 195 pages. 1996.

Vol. 1646: J.-P. Demailly, T. Peternell, G. Tian, A. N. Tyurin, Transcendental Methods in Algebraic Geometry. Cetraro, 1994. Editors: F. Catanese, C. Ciliberto. VII, 257 pages. 1996.

Vol. 1647: D. Dias, P. Le Barz, Configuration Spaces over Hilbert Schemes and Applications. VII. 143 pages. 1996.

Vol. 1648: R. Dobrushin, P. Groeneboom, M. Ledoux, Lectures on Probability Theory and Statistics. Editor: P. Bernard. VIII, 300 pages. 1996.

Vol. 1649: S. Kumar, G. Laumon, U. Stuhler, Vector Bundles on Curves – New Directions. Cetraro, 1995. Editor: M. S. Narasimhan. VII, 193 pages. 1997.

Vol. 1650: J. Wildeshaus, Realizations of Polylogarithms. XI, 343 pages. 1997.

Vol. 1651: M. Drmota, R. F. Tichy, Sequences, Discrepancies and Applications. XIII, 503 pages. 1997.

Vol. 1652: S. Todorcevic, Topics in Topology. VIII, 153 pages. 1997.

Vol. 1653: R. Benedetti, C. Petronio, Branched Standard Spines of 3-manifolds. VIII, 132 pages. 1997.

Vol. 1654: R. W. Ghrist, P. J. Holmes, M. C. Sullivan, Knots and Links in Three-Dimensional Flows. X, 208 pages. 1997.

Vol. 1655: J. Azéma, M. Emery, M. Yor (Eds.), Séminaire de Probabilités XXXI. VIII, 329 pages. 1997.

Vol. 1656: B. Biais, T. Björk, J. Cvitanic, N. El Karoui, E. Jouini, J. C. Rochet, Financial Mathematics. Bressanone, 1996. Editor: W. J. Runggaldier. VII, 316 pages. 1997.

Vol. 1657: H. Reimann, The semi-simple zeta function of quaternionic Shimura varieties. IX, 143 pages. 1997.

Vol. 1658: A. Pumarino, J. A. Rodrıguez, Coexistence and Persistence of Strange Attractors. VIII, 195 pages. 1997.

Vol. 1659: V, Kozlov, V. Maz'ya, Theory of a Higher-Order Sturm-Liouville Equation. XI, 140 pages. 1997.

Vol. 1660: M. Bardi, M. G. Crandall, L. C. Evans, H. M. Soner, P. E. Souganidis, Viscosity Solutions and Applications. Montecatini Terme, 1995. Editors: I. Capuzzo Dolcetta, P. L. Lions. IX, 259 pages. 1997.

Vol. 1661: A. Tralle, J. Oprea, Symplectic Manifolds with no Kähler Structure. VIII, 207 pages. 1997.

Vol. 1662: J. W. Rutter, Spaces of Homotopy Self-Equivalences – A Survey. IX, 170 pages. 1997.

Vol. 1663: Y. E. Karpeshina; Perturbation Theory for the Schrödinger Operator with a Periodic Potential. VII, 352 pages. 1997.

Vol. 1664: M. Väth, Ideal Spaces. V, 146 pages. 1997.

Vol. 1665: E. Giné, G. R. Grimmett, L. Saloff-Coste, Lectures on Probability Theory and Statistics 1996. Editor: P. Bernard. X, 424 pages, 1997.

Vol. 1666: M. van der Put, M. F. Singer, Galois Theory of Difference Equations. VII, 179 pages. 1997.

Vol. 1667: J. M. F. Castillo, M. González, Three-space Problems in Banach Space Theory. XII, 267 pages. 1997.

Vol. 1668: D. B. Dix, Large-Time Behavior of Solutions of Linear Dispersive Equations. XIV, 203 pages. 1997.

Vol. 1669: U. Kaiser, Link Theory in Manifolds. XIV, 167 pages. 1997.

Vol. 1670: J. W. Neuberger, Sobolev Gradients and Differential Equations. VIII, 150 pages. 1997.

Vol. 1671: S. Bouc, Green Functors and G-sets. VII, 342 pages. 1997.

Vol. 1672: S. Mandal, Projective Modules and Complete Intersections. VIII, 114 pages. 1997.

Vol. 1673: F. D. Grosshans, Algebraic Homogeneous Spaces and Invariant Theory. VI, 148 pages. 1997.

Vol. 1674: G. Klaas, C. R. Leedham-Green, W. Plesken, Linear Pro-p-Groups of Finite Width. VIII, 115 pages. 1997.

Vol. 1675: J. E. Yukich, Probability Theory of Classical Euclidean Optimization Problems. X, 152 pages. 1998.

Vol. 1676: P. Cembranos, J. Mendoza, Banach Spaces of Vector-Valued Functions. VIII, 118 pages. 1997.

Vol. 1677: N. Proskurin, Cubic Metaplectic Forms and Theta Functions. VIII, 196 pages. 1998.

Vol. 1678: O. Krupková, The Geometry of Ordinary Variational Equations. X, 251 pages. 1997.

Vol. 1679: K.-G. Grosse-Erdmann, The Blocking Technique. Weighted Mean Operators and Hardy's Inequality. IX, 114 pages. 1998.

Vol. 1680: K.-Z. Li, F. Oort, Moduli of Supersingular Abelian Varieties. V, 116 pages. 1998.

Vol. 1681: G. J. Wirsching, The Dynamical System Generated by the 3n+1 Function. VII, 158 pages. 1998.

Vol. 1682: H.-D. Alber, Materials with Memory. X, 166 pages. 1998.

Vol. 1683: A. Pomp, The Boundary-Domain Integral Method for Elliptic Systems. XVI, 163 pages. 1998.

Vol. 1684: C. A. Berenstein, P. F. Ebenfelt, S. G. Gindikin, S. Helgason, A. E. Tumanov, Integral Geometry, Radon Transforms and Complex Analysis. Firenze, 1996. Editors: E. Casadio Tarabusi, M. A. Picardello, G. Zampieri. VII, 160 pages. 1998

Vol. 1685: S. König, A. Zimmermann, Derived Equivalences for Group Rings. X, 146 pages. 1998.

Vol. 1686: J. Azéma, M. Émery, M. Ledoux, M. Yor (Eds.), Séminaire de Probabilités XXXII. VI, 440 pages. 1998.

Vol. 1687: F. Bornemann, Homogenization in Time of Singularly Perturbed Mechanical Systems. XII, 156 pages. 1998.

Vol. 1688: S. Assing, W. Schmidt, Continuous Strong Markov Processes in Dimension One. XII, 137 page. 1998.

Vol. 1689: W. Fulton, P. Pragacz, Schubert Varieties and Degeneracy Loci. XI, 148 pages. 1998.

Vol. 1690: M. T. Barlow, D. Nualart, Lectures on Probability Theory and Statistics. Editor: P. Bernard. VIII, 237 pages. 1998.

Vol. 1691: R. Bezrukavnikov, M. Finkelberg, V. Schechtman, Factorizable Sheaves and Quantum Groups. X, 282 pages. 1998.

Vol. 1692: T. M. W. Eyre, Quantum Stochastic Calculus and Representations of Lie Superalgebras. IX, 138 pages. 1998.

Vol. 1694: A. Braides, Approximation of Free-Discontinuity Problems. XI, 149 pages. 1998.

Vol. 1695: D. J. Hartfiel, Markov Set-Chains. VIII, 131 pages. 1998.

Vol. 1696: E. Bouscaren (Ed.): Model Theory and Algebraic Geometry. XV, 211 pages. 1998.

Vol. 1697: B. Cockburn, C. Johnson, C.-W. Shu, E. Tadmor, Advanced Numerical Approximation of Nonlinear Hyperbolic Equations. Cetraro, Italy, 1997. Editor: A. Quarteroni. VII, 390 pages. 1998.

Vol. 1698: M. Bhattacharjee, D. Macpherson, R. G. Möller, P. Neumann, Notes on Infinite Permutation Groups. XI, 202 pages. 1998.

Vol. 1699: A. Inoue, Tomita-Takesaki Theory in Algebras of Unbounded Operators. VIII, 241 pages. 1998.

Vol. 1700: W. A. Woyczyński, Burgers-KPZ Turbulence, XI, 318 pages. 1998.

Vol. 1701: Ti-Jun Xiao, J. Liang, The Cauchy Problem of Higher Order Abstract Differential Equations, XII, 302 pages. 1998.

Vol. 1702: J. Ma, J. Yong, Forward-Backward Stochastic Differential Equations and Their Applications. XIII, 270 pages. 1999.

Vol. 1703: R. M. Dudley, R. Norvaiša, Differentiability of Six Operators on Nonsmooth Functions and p-Variation. VIII, 272 pages. 1999.

Vol. 1704: H. Tamanoi, Elliptic Genera and Vertex Operator Super-Algebras. VI, 390 pages. 1999.

Vol. 1705: I. Nikolaev, E. Zhuzhoma, Flows in 2-dimensional Manifolds. XIX, 294 pages. 1999.

Vol. 1706: S. Yu. Pilyugin, Shadowing in Dynamical Systems. XVII, 271 pages. 1999.

Vol. 1707: R. Pytlak, Numerical Methods for Optimal Control Problems with State Constraints. XV, 215 pages. 1999.

Vol. 1708: K. Zuo, Representations of Fundamental Groups of Algebraic Varieties. VII, 139 pages. 1999.

Vol. 1709: J. Azéma, M. Émery, M. Ledoux, M. Yor (Eds), Séminaire de Probabilités XXXIII. VIII, 418 pages. 1999.

Vol. 1710: M. Koecher, The Minnesota Notes on Jordan Algebras and Their Applications. IX, 173 pages. 1999.

Vol. 1711: W. Ricker, Operator Algebras Generated by Commuting Projections: A Vector Measure Approach. XVII, 159 pages. 1999.

Vol. 1712: N. Schwartz, J. J. Madden, Semi-algebraic Function Rings and Reflectors of Partially Ordered Rings. XI, 279 pages. 1999.

Vol. 1713: F. Bethuel, G. Huisken, S. Müller, K. Steffen, Calculus of Variations and Geometric Evolution Problems. Cetraro, 1996. Editors: S. Hildebrandt, M. Struwe. VII, 293 pages. 1999.

Vol. 1714: O. Diekmann, R. Durrett, K. P. Hadeler, P. K. Maini, H. L. Smith, Mathematics Inspired by Biology. Martina Franca, 1997. Editors: V. Capasso, O. Diekmann. VII, 268 pages. 1999.

Vol. 1715: N. V. Krylov, M. Röckner, J. Zabczyk, Stochastic PDE's and Kolmogorov Equations in Infinite Dimensions. Cetraro, 1998. Editor: G. Da Prato. VIII, 239 pages. 1999.

Vol. 1716: J. Coates, R. Greenberg, K. A. Ribet, K. Rubin, Arithmetic Theory of Elliptic Curves. Cetraro, 1997. Editor: C. Viola. VIII, 260 pages. 1999.

Vol. 1717: J. Bertoin, F. Martinelli, Y. Peres, Lectures on Probability Theory and Statistics. Saint-Flour, 1997. Editor: P. Bernard. IX, 291 pages. 1999.

Vol. 1718: A. Eberle, Uniqueness and Non-Uniqueness of Semigroups Generated by Singular Diffusion Operators. VIII, 262 pages. 1999.

Vol. 1719: K. R. Meyer, Periodic Solutions of the N-Body Problem. IX, 144 pages. 1999.

Vol. 1720: D. Elworthy, Y. Le Jan, X-M. Li, On the Geometry of Diffusion Operators and Stochastic Flows. IV, 118 pages. 1999.

Vol. 1721: A. Iarrobino, V. Kanev, Power Sums, Gorenstein Algebras, and Determinantal Loci. XXVII, 345 pages. 1999.

Vol. 1722: R. McCutcheon, Elemental Methods in Ergodic Ramsey Theory. VI, 160 pages. 1999.

Vol. 1723: J. P. Croisille, C. Lebeau, Diffraction by an Immersed Elastic Wedge. VI, 134 pages. 1999.

Vol. 1724: V. N. Kolokoltsov, Semiclassical Analysis for Diffusions and Stochastic Processes. VIII, 347 pages. 2000.

Vol. 1725: D. A. Wolf-Gladrow, Lattice-Gas Cellular Automata and Lattice Boltzmann Models. IX, 308 pages. 2000.

Vol. 1726: V. Marić, Regular Variation and Differential Equations. X, 127 pages. 2000.

Vol. 1727: P. Kravanja, M. Van Barel, Computing the Zeros of Analytic Functions. VII, 111 pages. 2000.

Vol. 1728: K. Gatermann, Computer Algebra Methods for Equivariant Dynamical Systems. XV, 153 pages. 2000.

Vol. 1729: J. Azéma, M. Émery, M. Ledoux, M. Yor, Séminaire de Probabilités XXXIV. VI, 431 pages. 2000.

Vol. 1730: S. Graf, H. Luschgy, Foundations of Quantization for Probability Distributions. X, 230 pages. 2000.

Vol. 1731: T. Hsu, Quilts: Central Extensions, Braid Actions, and Finite Groups,. XII, 185 pages. 2000.

Vol. 1732: K. Keller, Invariant Factors, Julia Equivalences and the (Abstract) Mandelbrot Set. X, 206 pages. 2000.

Vol. 1733: K. Ritter, Average-Case Analysis of Numerical Problems. IX, 254 pages. 2000.

Vol. 1735: D. Yafaev, Scattering Theory: Some Old and New Problems. XVI, 169 pages. 2000.

Vol. 1736: B. O. Turesson, Nonlinear Potential Theory and Weighted Sobolev Spaces. XIV, 173 pages. 2000.

Vol. 1737: S. Wakabayashi, Classical Microlocal Analysis in the Space of Hyperfunctions. VIII, 367 pages. 2000.